Key Technologies for Tunnel Construction under Complex Geological and Environmental Conditions

Series Editor

Qihu Qian, Nanjing, China

This series is written on the basis of several difficult projects, aiming at complex geological conditions such as complex topography and landform types, complex geological structure and poor geotechnical engineering geology, systematically describes the Key Technologies for Tunnel Construction under Complex Geological and Environmental Conditions like mountains, underwater, soft and uneven strata, from the aspects of geological exploration, early warning and risk control, design, construction and digital application. It's a monograph that introduces innovative achievements and core technologies formed in the construction of complex geological tunnels in recent years. The series contains both texts and illustrations. It can be used as a reference for engineers, technicians, scientific researchers, teachers and students of related specialties in tunnels and underground engineering.

This series systematically describes the Key Technologies for Tunnel Construction under Complex Geological and Environmental Conditions like mountains, underwater, soft and uneven strata, from the aspects of geological exploration, early warning and risk control, design, construction and digital application.

Contents of the series are mainly from a number of key projects, such as Yuelongmen Tunnel of Chenglan Railway in Sichuan Province, Jinan Yellow River Tunnel Project, Qiyueshan Tunnel of West Hubei Expressway, Jiaozhou Bay Tunnel of Qingdao, Nanjing Yangtze River Tunnel and Wuhan River Crossing Tunnel, as well as the 973 Projects of the State-Mechanism, Prediction, Early Warning and Control of Major Water Inrush Disasters in Deep and Long Tunnels, Science and Technology Project by the Ministry of Transport "Study on Key Technologies of Underwater Large Section Shield Tunnel Construction under Complex Geological Conditions", National Defense Research Project "Three-Dimensional Advance Prediction Technology for National Defense Underground Engineering".

Rulu Wang · Dongmei Zhang · Chunsheng Guo · Jingya Yan

Structural Safety Early Warning and Treatment Technology of Metro Tunnel in Soft Soils

SSTP

Springer

Rulu Wang
Shanghai Shentong Metro CO., Ltd.
Shanghai, China

Dongmei Zhang
Tongji University
Shanghai, China

Chunsheng Guo
SGIDI Engineering Consulting (Group)
Co., Ltd.
Shanghai, China

Jingya Yan
Shanghai Shentong Metro CO., Ltd.
Shanghai, China

Translated by
Cheng Jiang
The College of Foreign Languages
Shanghai University for Science
and Technology
Shanghai, China

Yang Han
The College of Foreign Languages
Shanghai University for Science
and Technology
Shanghai, China

Shang Wu
The College of Foreign Languages
Shanghai University for Science
and Technology
Shanghai, China

Juyan Wu
The College of Foreign Languages
Shanghai University for Science
and Technology
Shanghai, China

ISSN 2662-2904 ISSN 2662-2912 (electronic)
Key Technologies for Tunnel Construction under Complex Geological and Environmental Conditions
ISBN 978-981-96-6409-2 ISBN 978-981-96-6410-8 (eBook)
https://doi.org/10.1007/978-981-96-6410-8

Jointly published with Shanghai Scientific and Technical Publishers
The print edition is not for sale in China (Mainland). Customers from China (Mainland) please order the print book from: Shanghai Scientific and Technical Publishers.

This Springer imprint is published by the registered company Springer Nature Singapore Pte Ltd.
The registered company address is: 152 Beach Road, #21-01/04 Gateway East, Singapore 189721, Singapore

Foreword

Since the late 1950s, Shanghai had been making attempts to build subway tunnels and river-crossing passages in the soft soils. However, the industrial foundation was weak and the technologies were inadequate at that time, resulting in a very slow progress in the work. Not until 1965, a shield tunnel with an external diameter of 5.8 m and a double-track length of 660m was built for a pilot test. And in 1971, the first tunnel crossing the Huangpu River at Dapu Road with an external diameter of 10.2 m was finally built, marking substantial progress in shield tunnel technology. The construction of the first river-crossing tunnel has laid a solid foundation for shield tunnel construction in soft soil areas. So far, a subway network of 705km in length has been built in Shanghai. It can transport 3.88 billion passengers per year, with an average daily passenger flow exceeding 12 million, accounting for more than 66% of public transportation. Shanghai Metro has become the most important part of urban transportation, and has been playing an increasingly important role in urban operation.

At present, there are about 41 cities in the mainland of China that have opened rail transit, with a total length of 6900km. And in quite a few cities the subway tunnels are built in soft soils, which has the characteristics of "low strength, high sensitivity and prone to large deformation". In addition, there are many joints in shield tunnels. Influenced by various factors, shield tunnels are prone to structural diseases such as water leakage, deformation, cracks of segment, and fragmentation, which will directly result in safety problems of tunnel structures. With the increasing extension of urban rail transit network, the public are getting more and more dependent on rail transit for their daily transportation. The daily operation service time of no less than 18 hours and high-level operation service requires higher standards for the maintenance of tunnel structure.

To address such issues as the unclear deformation developing mechanism of subway shield tunnel structure in soft soil areas, the inadequate inspection equipment and technology for early warning, the lack of performance assessment methods and pre-control protection standards for shield tunnel structure, and relatively backward technology and equipment for structural repairing and treatment inside and outside the tunnel, the writing team compiled the book *Early Warning, Pre-control and Treatment Technology of Metro Tunnel Structure Safety in Soft Soils* by sorting out the achievements in theoretical research, full-scale tests, equipment technology research and development, etc. This book expounds systematically and comprehensively the deformation mechanism of tunnels, intelligent sensing technology of structural damage, restoration technology of major structural damage of tunnels to lay a foundation for early warning, pre-control, and treatment of tunnel structure safety in soft soil areas. It is of great significance to provide guidance to the engineering constructions, and intelligent operation and maintenance of urban rail transit tunnels.

The compiling team of this book have been committed to subway construction, operation, and maintenance for 20–30 years. And they have made many innovative explorations, accumulated rich practical experiences, and made many excellent research achievements in the field of tunnel operation and maintenance in soft soil areas. Thank you for your presentation without reservation in this book. I hope that the book will attract the attention of industry experts and scholars so that we can work together to improve the theoretical and practical methods, and play a greater role in ensuring the structural safety of subway tunnels in soft soil areas.

Beijing, China
January 2021

Wenhao Liang

Preface

The safety of subway operation in large cities and super-large cities has become an important part of the safe operation of urban lifeline, while the safety of subway tunnel structure serves the key foundation for subway safe operation. On January 7, 2018, *Opinions on Promoting Urban Safety Development* issued by the State Council proposed that urban safety assurance capabilities, the innovation and application of security technology be strengthened, the investment in urban operation facilities be increased, the advanced production technique and safety technology be promoted, automatic monitoring and pre-control capabilities be enhanced, the informatization construction of urban safety supervision be strengthened, the intelligent system for urban safety management be accelerated, the construction of urban lifeline be promoted, and advanced safety technologies and products such as risk and disaster prevention and control, monitoring and early warning, individual protection, emergency response, and earthquake engineering be developed, promoted, and applied.

In response to "the Opinions", this book summarizes the practical experiences and the research results of subway operation and maintenance in Shanghai soft soil areas in the recent 20 years, aiming to improve the theoretical level, control technology and management efficiency of tunnel structure safety in domestic urban rail transit operation. It expounds systematically the pre-control and treatment technology of subway tunnels in the operation process, focusing on the tunnel deformation evolution mechanism based on a large number of operation and maintenance data and engineering practices. The safety control specifications of tunnel structure is proposed by the destructive test of tunnel lining and the reinforced segments. It also introduces the structural reinforcement construction technology and special maintenance and treatment equipment that can be utilized outside and inside the tunnel. By integrating the collected data related to tunnel structural safety by means of information technology, it provides a high-efficiency, whole-process and full-coverage information management platform to solve the problems of detecting, assessing, and treating the hidden safety hazards in the tunnel structure so as to ensure the long-term safe and stable operation of urban rail transit. The book can be used for reference by researchers and practitioners in the related fields.

This book is divided into eight chapters. Chapter 1 is the overview. Chapter 2 describes the deformation mechanism of tunnel structure, Chap. 3 discusses the intelligent sensing technology of tunnel, Chap. 4 analyzes the ultimate bearing capacity of subway tunnel structure and the research on bearing capacity after reinforcement, Chap. 5 deals with the deformation damage repair technology of subway tunnel structure, Chap. 6 discusses the deformation treatment technology of double-liquid micro-disturbance grouting, Chap. 7 proposes an intelligent operation and maintenance management system of tunnel life cycle, and Chap. 8 discusses the engineering application. This book is co-authored by Rulu Wang, Dongmei Zhang, Chunsheng Guo and Jingya Yan. Chapter 1 is written by Rulu Wang and Lujie Wang; Chaps. 2, 4, 5 and 6 are written by Rulu Wang, Dongmei Zhang, and Jingya Yan. Chapters 3 and 7 are written by Chunsheng Guo, Jiaping Li, and Zhao Yuan. Chapter 8 is written by Jingya Yan and Tang Tao. In the process of writing, we have received guidance and help from many senior experts, scholars, and colleagues in the industry. We would like to express our sincere gratitude for their support and help.

We would like to express our gratitude to Mr. Liang Wenhao, the academician from the Chinese Academy of Engineering for writing the preface to this book, and his constant guidance and help in the writing process.

We would like to thank China Railway Publishing House Co., Ltd. for its strong support and hard work in publishing this book.

Due to the limited time and academic level, there may be a few errors in the book. And some opinions may be immature, which need further tests in more engineering practices. All the suggestions from the readers are welcome.

Shanghai, China
March 2021

Rulu Wang
Dongmei Zhang
Chunsheng Guo
Jingya Yan

Contents

Chapter 1
Introduction

1.1 Introduction

By the end of December 2020, a total of 45 cities in the Mainland of China had opened urban rail transportation systems, with an operating mileage of 7978.19 km, as shown in Fig. 1.1. Such large cities as Beijing, Shanghai, Guangzhou, Shenzhen, Nanjing, Wuhan, and Chengdu have basically realized "one-ticket transfer" operation in the "interconnection and interchange" urban rail networks. And another more 50 cities, according to the planning and construction arrangement, are constructing, or preparing for the construction of urban rail transit projects, the operation mileage of lines reaching up to nearly 8000 km in total.

By the end of March 2021, the total operation mileage of Shanghai Metro network would have increased to 772 km, with 459 stations in total. In 2020, Shanghai Metro network transported 2.835 billion passengers. The average daily passenger flow on weekdays reached up to 7.746 million, the peak flow being 12.3671 million trips, accounting for over 66% of public transportation. According to the construction plan, another 61 km of subway would be put into operation in 2021, and by then a rail transit network of 774.6 km would be formed, and the proportion of passenger flow in Shanghai's urban public transportation would be more than 70%, and Shanghai Metro has been playing an increasingly important role in the urban public transportation, as shown in Fig. 1.2. The proportion of metro passenger flow having been on the rise, people are showing their deep concern over the safety of metro operation. The malfunction of Shanghai rail transit during the rush hours is bound to result in "butterfly effect", which would affect the public transportation of the whole city. Therefore, the safety of subway operation has constituted an essential part of the lifeline projects for large cities and megalopolis, while the safety of tunnel structure serves as the solid foundation for the safe operation of subways.

R. Wang et al., *Structural Safety Early Warning and Treatment Technology of Metro Tunnel in Soft Soils*, Key Technologies for Tunnel Construction under Complex Geological and Environmental Conditions,
https://doi.org/10.1007/978-981-96-6410-8_1

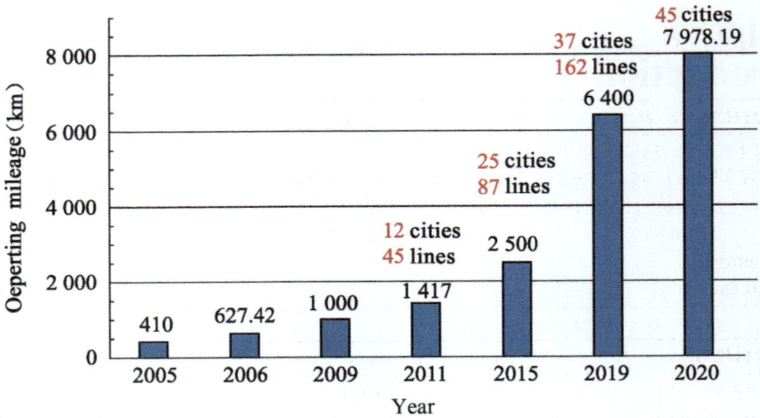

Fig. 1.1 National rail transit distribution bar chart

However, a large number of subway tunnels in coastal cities or cities on rivers and lakes are built in soft soil strata, and many subway tunnels are even built in mucky soil stratum. Obviously, this kind of soft soil, in general, has the characteristics of "low strength, high sensitivity and easy to deform". Once the ground is disturbed, the soil deformation is large, the strength decreases significantly. Moreover, the deformation can last for quite a long time. Since there are many joints in the shield tunnels, and the environment inside and outside the tunnel is variable and complicated, in addition to the influence of many other factors, the tunnel structure is prone to water leakage, large deformation, segment crack and crushing, and other serious structural damages, which directly affect the safety of the tunnel structure. And it is extremely difficult to repair the tunnel lining once it has been damaged (Liu and Hou 1991). Such accidents in metro tunnels had occurred in many cities at home and abroad. Take for example, in December 1995, a large amount of water and sand flooded into the tunnel between the "Forest" (JlecHafl) and "Muzhestva Square" (Russian: iLiowaflb MyacecTBa) stations of the St. Petersburg Metro Line 1 (Red Line). The damage was beyond control and the tunnel had to be buried up by refilling water into it, resulting in the permanent closure and abandonment of the section of rails between the two stations after 20 years of operation. This accident resulted in the closure of this section of subway, the worst of all, it had severely affected the urban operation. Due to the closure and abandonment of the tunnel, turn back lines at two stations had to be built for trains to turn back. Since then, the Red Line had been cut into two sections, with more than 500,000 regular passengers per day having to transfer to free buses at either Forest or Muzhestva Square station to cross the closed subway section so that they can continue their remaining journey on the other side of the subway line, which has increased their travel time and greatly affected the efficiency of transportation. The reconstruction for this rail section (double lanes 800 m in length) had to make a detour, which was completed and opened in June 2004. It took more than eight years and cost $145 million to complete the reconstruction project. Similarly, it had been

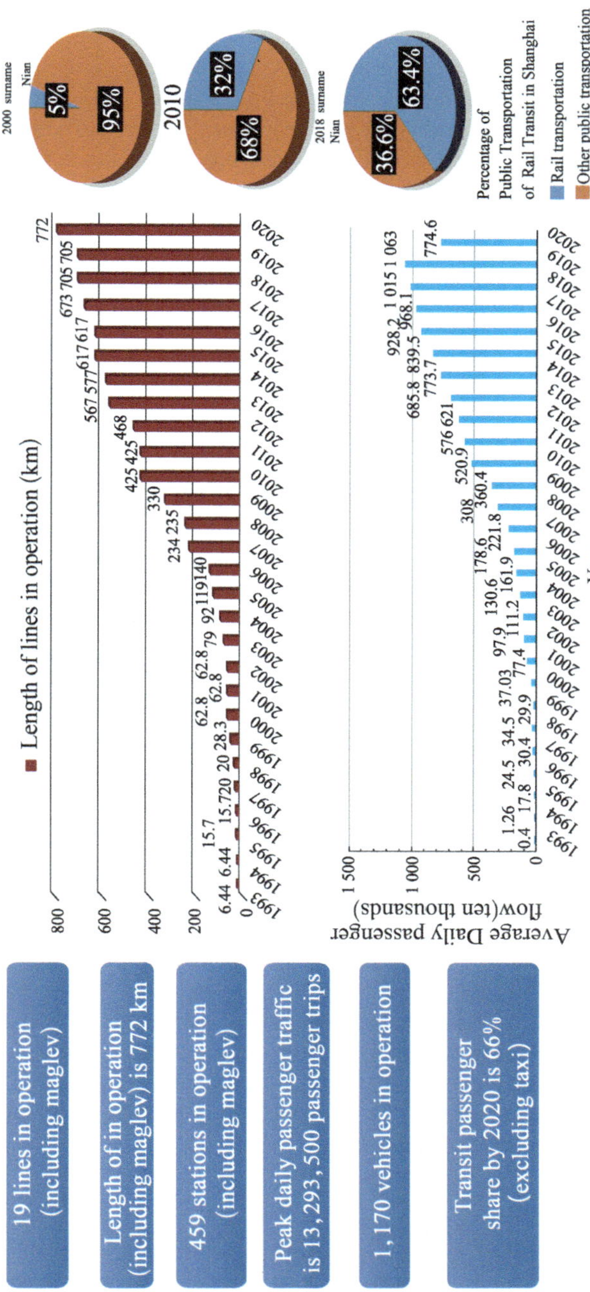

Fig. 1.2 Shanghai Metro development

reported abroad the cases of continuous settlement for years in the underlying strata when shield tunneling was completed (Hu et al. 2008).

The past two decades has witnessed the construction highlights of China's urban rail transit. However, due to the lack of experience in planning, design, construction and operation, as well as the relatively weak technical forces and equipment, safety accidents and dangers had occurred in recent years during the construction of tunnels or in the tunnels already in service operation. Subway tunnels in soft soil areas in particular are prone to significant deformation due to the complicated and variable environmental conditions. Figure 1.3 shows the settlement curve of Shanghai Metro Line 1. Excessive deformation can lead to tunnel seepage and leakage, structural cracks and other serious structural diseases. Such cases had occurred in Huangpi South Road section of Shanghai Metro Line 1, the East Extension of Chuangxin Central Road to Huaxia East Road section of Line 2, Qihua Road to Gucun Park Station section of Line 7, Shipyard Road section of Line 7, Shanghai Hongqiao section Line 2 etc., which had affected the subway operation. Shown in Fig. 1.4 are the types of diseases of shield tunnel structure.

What's more, in more than ten cities such as Nanjing, Guangzhou, Hangzhou, Shenzhen, Ningbo, Zhengzhou, Wuhan, Kunming, Harbin, Xiamen, Fuzhou, Suzhou etc., diseases of the tunnel structure like large deformation, tunnel lining crack, compression and expansion of the joints, joint dislocation, water and mud leakage, sand gushing, and damage to the waterproof sealing gaskets and bolts had occurred in the constructing tunnels or operating tunnels. Shown in Fig. 1.5 are the work sites of reinforcement and treatment of the damaged tunnel structures. It is quite difficult, costly, and time-consuming to repair the seriously damaged tunnel structures, if once they had occurred. As result, the subway operations may even have to be suspended due to the repairing work, causing major disturbance to the urban operation.

Structural damage imposes great threat to the safety of metro tunnels. The causes of structural diseases of tunnels are so variable and complicated that it is extremely difficult to repair them once the damage occurred.

Therefore, in the complicated construction environment, it is essential to learn about the trend of deformation, explore the deformation mechanism so that an effective smart management system for the tunnel networks can be established to ensure that risks in tunnel structure can be detected, dragonized, and controlled timely and quickly.

It is proposed that the "Inspection, Evaluation and Repair" (IER) process be followed in order to identify structural diseases in time and adopt the correct and appropriate maintenance methods or approaches so as to curb, eliminate, or control the deterioration of structural diseases, and ensure the long-term structural safety, the functionality, and durability of the tunnels.

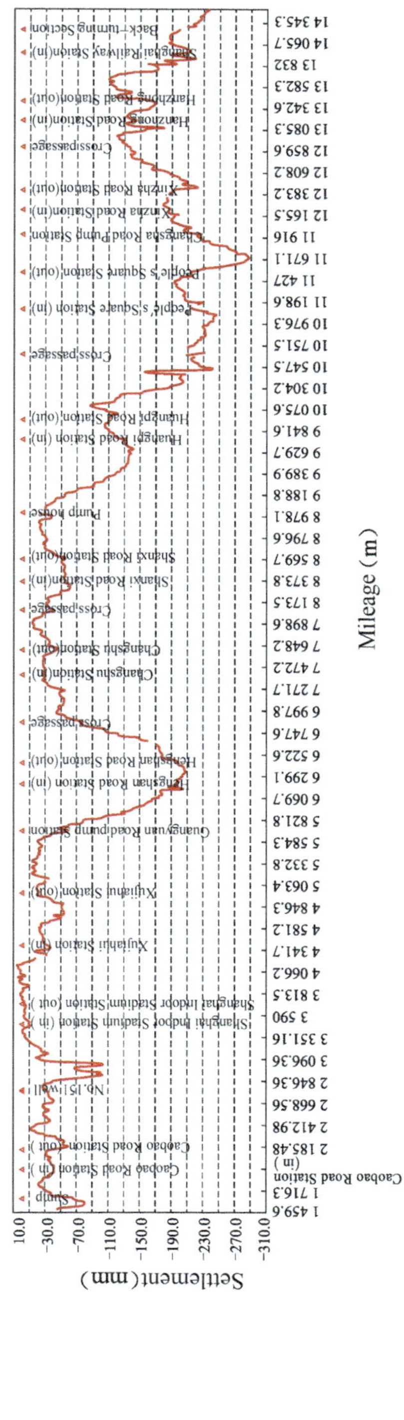

Fig. 1.3 Cumulative settlement curve of Shanghai Metro Line 1 (1995. 6 ~ 2017. 12 old points)

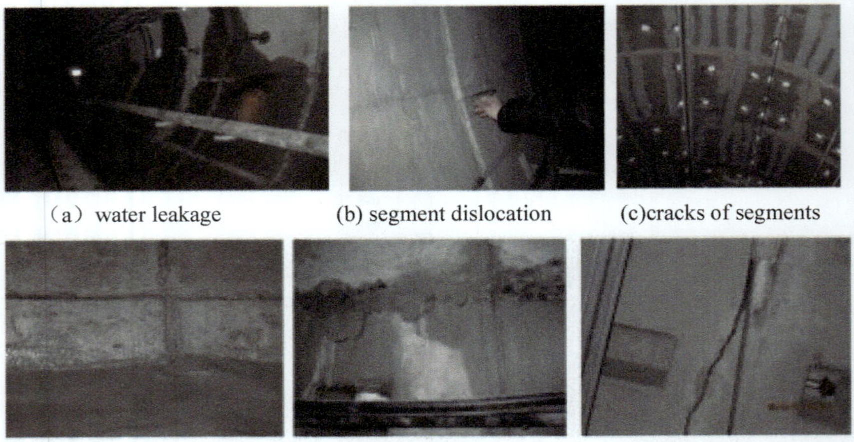

(a) water leakage (b) segment dislocation (c)cracks of segments

(d) mud and sand leakage (e) compression damage to segments (f) tensile failure of bolts

Fig. 1.4 Structural damage to the tunnel lining

(a) plugging by grouting in tunnel (b) Temporary support in (c) Pressure-weighted plugging in tunnel

Fig. 1.5 Tunnel reinforcement and defects treatment

1.2 Current Status of Tunnel Structure Safety Assurance Domestically and Internationally

1.2.1 State of Tunnel Monitoring, Detection, and Maintenance Out of China

The detection and treatment of structural diseases, by means of technology and equipment for timely detection, quick diagnosis, and alerting, serves as the core of the safety guarantee system for tunnel structures. It is required that the damaged structure be maintained and repaired when it reaches the standard for maintenance, in which the aids of technology and equipment play an important role.

Theoretical studies and the development of technical equipment for the safety protection of subway tunnels in foreign countries began much earlier than that of

China. The Maintenance and Repair of Underground Structures of ITA, an authoritative international tunnel research institution, published its research reports successively *Civil Engineering Records of Underground Structures* in 1987, *Report on-the Damaging Effects of Water on Tunnels During Their Working Life* in 1991, and *Testing Methods for Determining State of a Tunnel Lining* in 1995. Since 1997, the USA had conducted a systematic study on the detection, maintenance and repairing of tunnels. In 2003, a technical manual and guideline for the safety management of tunnels in operation service was officially released. In Europe, more research had been made to improve the operational reliability, reparability, and safety of tunnel structures. In recent years, since several accidents that jeopardized the safety of railroad operation tunnels had occurred in Japan, the Ministry of Construction of Japan made a survey of all the 3529 road tunnels across the country. Statistics show that structural diseases had occurred in more than 60% of the tunnels. The survey had yielded the *Handbook of Road Tunnels Aging in Japan* and the *Handbook of Tunnel Reinforcement and Restoration.* With the constant development of computer technology and non-destructive testing technology, a visualized approach for underground tunnels inspection has been gradually developed abroad. The KARO system developed in Germany in the 1990s and the Pipe Inspection Real-time Assessment Technique (PIRAT) system in Australia adopted this approach. Afterward, a multi-camera Sewer Scanner and Evaluation Technology (SSET) system was developed in the USA, utilizing multiple cameras to improve effectively the recording accuracy for image acquisition. In 2000, a Japanese company also employed multiple cameras for image acquisition. The system recorded images during the inspection process, and it usually took a long time for the operators to identify the diseases afterward. Recently, it has been significantly improved to conduct automatic identification and analysis of the diseases from the collected images. The automatic identification technology using camera for detecting tunnel diseases has been used worldwide. Although it could carry out rapid on-site information acquisition and automated identification of obvious diseases, yet the limited identification accuracy was far from satisfaction because it could only detect large scope of wet stain, while the smaller diseases such as cracks of segments could not be detected.

In the 1960s, steel lining reinforcement was applied in South Africa and France. The researchers adhered the steel plate to the concrete tensile surface by epoxy resin to enhance the load-bearing capacity of reinforced concrete. The application of this method was gradually promoted due to its good effect around the world.

1.2.2 State of Monitoring, Testing, and Maintenance in China

In China, relevant research work had been conducted on the detection technology and assessment method for railroad tunnels. *Regulations for Non-destructive Testing of Railway Tunnel Lining Quality* take for example was issued by the Ministry of Railways in 2004, proposing the technical norms and procedures for the detection of concrete lining quality of railroad tunnels by geo-radar and acoustic wave methods. In 2004, Shanghai Metro formulated *Regulations for Shanghai Metro Shield Tunnel Maintenance.* And *Technical Regulations for Cross-River Tunnels Maintenance* was released with the guidance from Shanghai Municipal Construction and Transportation Commission.

At present, the inspection of the structural safety of domestic urban rail transit tunnels mainly focuses on the apparent inspection and measurement of the inner wall of the tunnel, with occasional involvement in the inspection of the internal structure and wall thickness. Apparent inspection is mainly made by means of traditional manual monitoring with the aids of automated equipment. 3D laser scanning, video, and photo-taking had been gradually utilized in some cities. Manual optical equipment had been mainly used for the monitoring, and also fully automated high-precision equipment had been used. Due to the specific environment, working in tunnels has the following characteristics:

(1) All types of work in subway tunnels are confined in a "narrow, long" space.
(2) The work "window time" is very short, and the requirements are high. The work in the tunnel has to be completed within the window time in order not to affect the normal operation of the subway the next day, the window time for working is only 2 ~ 3 h. So the traditional manual inspection mode, due to its low efficiency, can no longer satisfy the needs of the development of urban rail transit network. Moreover, the conventional monitoring is conducted every 3–6 months. Therefore, many structural deformations at an early stage cannot be detected in time. And the deformations and damages have already been very serious by the time they are found, making it quite difficult to carry out repairing work at a later stage. For this reason, it is quite necessary that the inspection equipment of "automated high-precision, large-range, high-efficiency" be developed and utilized for shield tunnels.

1.2.2.1 State Maintenance Technology and Equipment

In recent years, the structural diseases have been effectively treated in a number of tunnels by means of the existing tunnel reinforcement and repairing technologies. However, the tunnel reinforcement technologies available including waterproofing materials, the design and manufacture of special tools, structural reinforcement, foundation treatment, waterproofing construction, and other technical fields are lack of standardized technical regulations. Especially, for some special cases, it is not appropriate to follow up the existing specifications in terms of quality and safety control. And the reinforcement work is inefficient and the results are far from satisfaction. What's more, there is no follow-up inspection and assessment of their reinforcement effects. So far specialized technical equipment for proactive intervention inside and outside the tunnels are not available when deformation occurs. There is a lack of specialized large-scale equipment for structural management of operational tunnels due to the limited space inside the tunnels with the numerous pipelines and other facilities. And it is impossible to utilize large construction equipment, or to move heavier units or components in the confined space of tunnels as that on the ground surface, resulting low work efficiency. Therefore, it is essential to develop highly efficient equipment for confined spaces and proactive intervention equipment and technology for structural maintenance inside and outside tunnels.

Water seepage, cracking, and deformation of the tunnel structure resulting from such factors as water leakage in the tunnel, surcharge, and disturbance of the construction activities in the neighborhood are inevitable. Although the structural safety assessing standards for tunnels in operation have been initially formed after more than 20 years of practice in the Shanghai Metro, yet in terms of standard specifications and system, the knowledge about the development law of structural performance under multiple working conditions is inadequate. And it is impossible to determine what level of deformation or damage to the tunnel is acceptable for operational safety. Without the support from the full-scale tests and objective evaluation, no conclusion can be drawn from the previous work in design, construction, operation, and maintenance. Because it is difficult to judge whether the tunnel structures are safe or not by the results of the tests, the timing of structural reinforcement has to be determined on the basis of previous experiences. And there is a lack of theoretical basis about the countermeasures. As the stress conditions and failure mechanism of tunnel lining structures are very complicated, the key points of structural safety control lie in establishing the theoretical basis and methods for investigating the mechanism and ultimate bearing capacity of metro tunnel lining structures, assessing structural safety, and decision-making for maintenance and reinforcement. Therefore, it is urgent that a large number of experimental studies in different operating conditions be conducted to obtain objective criteria for safety assessment so that the optimal timing for structural reinforcement can be determined.

1.2.2.2 State of Maintenance Management

The increase of detection efficiency has resulted in a geometric growth in the number of subway tunnels that need maintaining. The variety of data types and the inconvenience of data application inevitably lead to pressure on tunnel maintenance. The key to the safe operation of tunnels lies in how to make a rational deployment of resources for tunnel maintenance. Therefore, the assessment and classification of the structural safety levels of the tunnel structure should be based on the results of the tunnel inspections, taking into account the geological and construction data, so as to make plans, and allocate different maintenance forces to ensure a balanced and timely treatment of the hidden hazards in tunnel structure safety at the network level. As the number of metro lines in service operation has been increasing in Shanghai, the lack of an advanced and reliable operation and maintenance management system for processing the big data collected from the geological information, monitoring and inspection, diseases treatment, structural reinforcement etc. makes it quite difficult to make rational deployment of the limited resources for operation and maintenance. How to improve the efficiency and effectiveness of maintenance, to effectively utilize the limited resources, and strength to achieve the best results with the minimum cost constitutes a major challenge for the management and maintenance of large and complicated tunnel damages and defects. In addition, it is necessary to control the current status and development trends of structural damages via the network so as to achieve full coverage and control of the whole process. It is essential to develop an intelligent operation and maintenance system and a digital information platform for Shanghai Metro to improve management efficiency and optimize the allocation of resources. By means of information technology, the collected data of the tunnel construction, hydrogeological conditions, 3D coordinates of the engineering route structure, the surrounding environment of the project, structural settlement and deformation, tunnel detection, water leakage, and the historical information of maintenance and management can be pooled and integrated so as to improve the efficiency of statistical analysis. The establishment of a practical digital information system can not only address the issues of data cleaning, filtering, storage, and effective application of massive data, but also help allocate the limited maintenance resources to facilitate the management of tunnel structural damages and defects. In addition, it can also help to set up a dynamic feedback mechanism by integrating the detection results, evaluation standard system, and tunnel maintenance and management. A supporting management system and work requirements can be formed on this basis to guarantee the long-term safety and stability of the tunnel structures of the urban rail transit in service.

1.3 Latest Achievements in Tunnel Maintenance in Shanghai

But how can we ensure the structural safety of metro shield tunnels? Such problems as unknown deformation evolution mechanism of tunnels, inaccurate and untimely detection and warning of structural defects, the lack of standards for structural safety assessment and pre-control, and the lack of restoration and treatment technologies, and the backward intelligent management for operation and maintenance have constituted a big challenge for Shanghai metro and Metro management units all over the country.

After more than 20 years of continuous research on the structural safety of shield tunnels in the soft soil in Shanghai, a set of solutions from theoretical analysis to equipment and technology to the structural safety of tunnels has been proposed (Fig. 1.6), which can solve the following five key problems.

1. Analysis of structural deformation of shield tunnels in soft soil area. Based on the long-term monitoring of the structural performance development of tunnels in service, the failure and damage modes of the segments in operation service are proposed, and the key diseases-inducing factors that affect the structural bearing capacity of the tunnel segment are analyzed; a model for analysis of the structural mechanics of the tunnel structure in the soft soil stratum is constructed; based

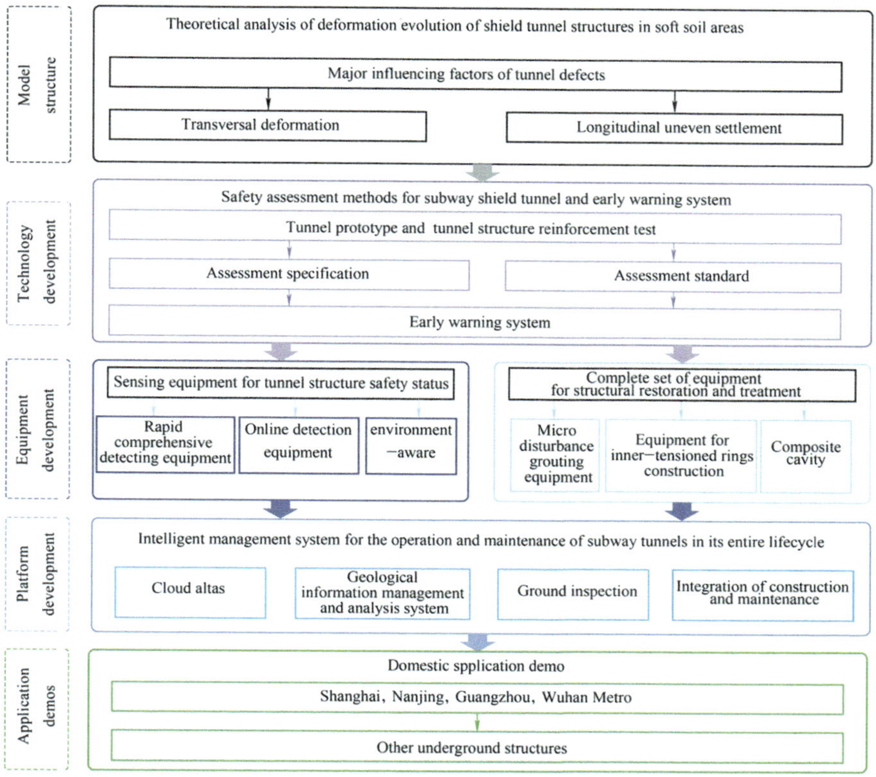

Fig. 1.6 Scheme for structural safety of tunnels

on the data of the shield disease monitoring and the test results of the tunnel structure, the mechanism of the shield tunnel structural failure is analyzed; and the evolution mechanism of structural deformation of the tunnel structure in the soft soil areas is investigated.

2. The development of an intelligent detection and rapid warning system for the structural safety state for multi-application scenarios and multi-defects identification. Based on the research of tunnel structural failure and its apparent phenomena, and considering the internal environmental conditions of the tunnels, a set of system for automatic and rapid collection and identification of tunnel damage is established by means of rapid image recognition technology; a method and equipment for detecting the differential settlement of subway tunnels in service are developed, realizing continuous monitoring of tunnel settlement and calculation of the settlement model; A method for detecting the convergence of the cross section of operating subway tunnels is proposed; Based on the camera measurement method and by means of an industrial control-level video system, the collection of effective information of the tunnel contour in motion is realized,

and the difficult technical problems of repeated positioning of the measurement section are solved; an online monitoring technology for the operating subway tunnel structure is developed; and a vehicle for comprehensive inspection of the operating subway tunnel structure is developed.

3. Assessment method for shield tunnel structure service performance and pre-control protection standard. Full-scale test is adopted to test the performance of circular tunnel segments. By simulating the actual stresses on the tunnel segments under unfavorable working conditions, the development of the ultimate stress, and the development law of ultimate deformation capacity and opening of joints, etc. is explored. The control specifications for the cross-sectional deformation of the tunnel segments are proposed on the basis of the tunnel structure and its normal state in operation service; a safety assessment and early warning system is built on the basis of the data collected from the on-site inspection of the tunnel structure.
4. The development of a complete set of processes and equipment for structural restoration and treatment that can be used both inside and outside tunnels. It is usually difficult to control the direction of slurry flow in the current grouting construction process; a low-pressure multi-directional injection grouting technology is developed independently. And a new grouting reinforcement process is developed for existing tunnels to realize uniform grouting around the tunnel lining, which not only improve the waterproof capacity, but also strengthens the surrounding soil body so as to improve the lateral resistance of tunnels, and decrease their deformation. A special manipulators and equipment platforms for the installation of steel plate and the reinforcement process for the steel plate of existing tunnels have been developed. Regulations and technical standards for tunnel reinforcement and restoration methods have been formed to achieve standardized construction and management, improve the quality of reinforcement and restoration works, and realize precise risk control.
5. The development of an integrated intelligent operation and maintenance management system for the whole-life cycle of metro tunnels to realize efficient management of the structural safety of the huge operation network, providing novel ideas for the structural safety of the subway shield tunnels. Digital management technology is developed for the processing the traffic data to achieve an efficient integrated management of the data from the survey, design, construction, monitoring, and surrounding environment, etc. Based on the collected data from the whole-life span and assessment technology, and by integrating the defect information, treatment plan and economic factors, the technical guide for the precautionary maintenance of tunnels is established. The integrated technology for construction and maintenance of urban rail transit is developed. Historical records for the diseases of the subway tunnel structures are established for the structural condition analysis and assessment so as to build an intelligent operation and maintenance management system for the tunnels in its whole-life cycle.

1.4 State of Intelligent Operation and Maintenance of Tunnels

Through theoretical research, development of intelligent detecting equipment, formulation of specialized maintenance equipment and technologies for tunnels, and development of an information-based maintenance management platform, defects can be promptly identified and controlled, thereby enhancing the level of intelligence and smartness in the operation, maintenance, and repair of tunnel structures.

1. Further theoretical study of structural safety of tunnels. For the safety of tunnel structures in different installation environment and geological conditions, intensive research on the evolution mechanism of structural deformation can be conducted to describe and reveal accurately the evolution mechanism of structural deformation of tunnels. Research can be made on the safety evaluation methods, indicator systems and standards for the tunnel structures in service operation.
2. Research on intelligent tunnel sensing systems. The use of sensing equipment such as 3D laser scanning, raster fiber optic sensing and infrared sensing will become normal for the detection of various types of damage to the tunnel structure. Monitoring and inspection equipment of artificial intelligence with autonomous learning will be developed to provide real-time sensing of the structural state of tunnels, timely prediction, prognosis and early warning of structural safety conditions so that proactive interventions can be realized to achieve pre-control and treatment of the tunnel structural defects.
3. Research on the integration of high-tech and social management. At present, satellites, planes, and drones can be used to identify the ground elevation changes automatically and give an early warning. However, the all-weather monitoring task cannot be realized some time due to weather conditions. Under this circumstances, regular inspections of the key areas can be conducted by integrating social grid management to eliminate hazards to the tunnels resulted from unexpected accidents.
4. By adopting big data, internet, Internet of Things (IoT), knowledge graph, artificial intelligence, and deep learning technology, a deep integrated analysis of the surface, strata, and the external and internal factors influencing the tunnel structure can be conducted to establish the strong and weak correlations between these factors so that reliable qualitative and quantitative prediction, pre-judgment, and warning can be made.
5. Research on the influence of various external construction operations on tunnel structures can be conducted by means of artificial intelligence technology to generate relevant report files automatically for the accurate prediction of the variation trends, and timely warnings.
6. Intelligent sensing components and equipment of small size, low energy consumption, multi-functionality, long lifespan, and low interference will be developed for real-time monitoring of tunnel structures and environments.

7. Research on active intervening techniques for tunnel structures and reinforcement construction equipment technology that can be conducted inside and outside the tunnels to form a systematic series of specialized methods for tunnel maintenance and curing.
8. Research and development of high-precision monitoring devices that can move quickly inside the tunnel to enable timely monitoring or trouble shooting.
9. Research and development of flexible, fast, and small transportation equipment that can transport a small amount of tools and personnel to the work site quickly.
10. Development of specialized equipment that is highly automated, efficient, durable, and easy to operate for the maintenance and management of tunnels.
11. Development of small and flexible equipment and tools, such as specialized auxiliary devices for efficient drilling, cutting, and automatic welding on steel plates.
12. Development of emergency response equipment for tunnels, such as devices for dealing with water leakage and surging sand, and rapid tunnel lifting.
13. Application of automatic monitoring and early warning of surface variation along subway lines, and the regular application of such new technologies as remote sensing by drones, satellites, and airplanes to know the surface changes and related special operations in time, and strengthen the automatic monitoring of groundwater along the subway lines as well.
14. Development of intelligent platforms for the operation and maintenance management of tunnels in its lifecycle. By means of information technology, the big data related to tunnel structure safety can be collected and integrated to provide an efficient, comprehensive end-to-end support for tunnel structure safety management to realize the goals that the status of tunnel structures can be known, safety risks can be controlled, and results can be expected, reaching a higher level of intelligence management.
15. Strengthening research on the handling of tunnel water seepage. Water seepage in tunnels is specific and it generally does not damage the existing waterproof system. Research on high-quality grouting techniques that do not damage the existing waterproof system can be conducted to ensure the grouting materials used should have long-term durability.
16. Enhancing research on the impact of stray currents on tunnel structure durability.
17. Research on the technology and equipment for replacement of a whole tunnel segment.

References

Hu X, Bai N, Li H (2008) Analysis of the tunnel accident in Saint Petersburg Metro Line 1. Tunnel Constr 28(4):5

Liu J, Hou X (1991) Shield tunneling method. China Railway Publishing House, Beijing

Chapter 2
Deformation Mechanism of the Tunnel Lining

2.1 Diseases of Shield Tunnel Lining

2.1.1 Types of Tunnel Lining Diseases

Common structural diseases of shield tunnels in soft strata include water leakage, structural damage, and tunnel deformation. There is a causal link between them. Failure to treat them in time will result in deterioration of structural performance, forming a vicious cycle. A typical case is the interaction between water leakage from tunnels and structural deformation, which makes the condition of water leakage and structural deformation even worse. Shown in Table 2.1 are the types of structural diseases of shield tunnels.

2.1.2 Water Seepage

Water leakage, the most common type of tunnel diseases, constitutes a pervasive challenge for shield tunnels worldwide. In terms of sectional distribution, it mainly occurs in such locations as portal shafts, connected aisles and pumping stations, small radius curves, poor geological conditions, and accident-prone locations. In terms of spatial distribution, it mainly occurs in circumferential and longitudinal joints, hand holes, bolt holes, grouting holes, and cracks of segments. Apart from lifting holes and cracks, almost all water leakage occurs at tunnel joints. Statistics show over 93% of water leakage occurs at joints of shield tunnels in Shanghai. Therefore, water leakage control at joints plays a key role in the prevention and treatment of structural diseases in tunnels. The distribution of tunnel water leakage is shown in Fig. 2.1.

R. Wang et al., *Structural Safety Early Warning and Treatment Technology of Metro Tunnel in Soft Soils*, Key Technologies for Tunnel Construction under Complex Geological and Environmental Conditions,
https://doi.org/10.1007/978-981-96-6410-8_2

Table 2.1 Types of structural diseases of shield tunnels

Types	Description
Water seepage/leakage	Wet stains, water seepage, water leakage, mud leakage, sand leakage
Structural damages (damages to components)	Cracks and fragmentation of tunnel segments Corner spalling, edge spalling, and block spalling Shear of bolts Corrosion of the metal connectors Gasket damage (by pulling, tearing, etc.) Joints dislocation (between rings, between segments) Open joints (mainly occurring in longitudinal joints) Excessive compression damage at joints (mainly occurring in longitudinal joints) Track bed disengagement (crack between track bed and tunnel segments)
Tunnel deformation	Differential longitudinal settlement Transverse convergence

2.1.3 Lining Damages

Lining diseases in various components and segments of the shield tunnels primarily manifests as cracks, fragmentation, corner spalling, edge spalling, block spalling, compression damage to segments, shear of bolts, corrosion of metal connectors, gasket damage, compression damage due to excessive pressure at joints, and track bed disengagement, etc. The distribution of structural diseases is shown in Fig. 2.2.

2.1.4 Differential Longitudinal Settlement of the Tunnel

When put into operation, the tunnel is prone to differential longitudinal settlement due to such influencing factors as the ground conditions, environmental conditions, the quality of tunnel construction and maintenance. Figure 2.3 shows the differential longitudinal settlement of the 16.1 km long line Phase 1 of Shanghai Metro Line 1 after 22 years of operation service. Differential settlement in the longitudinal direction will not pose a significant risk to shield tunnel if it meets the requirements for structural safety, waterproofing etc. However, special attention should be given to it if the deformation continues to develop and amounts to the threshold that structural safety is jeopardized.

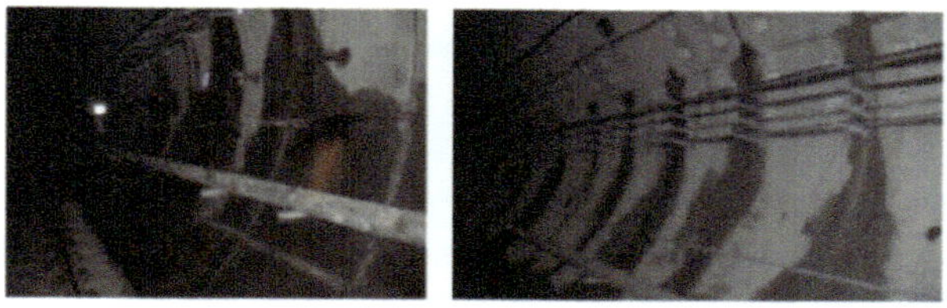

(a) Water leakage in circumferential joints

(b)Water leakage in longitudinal joints

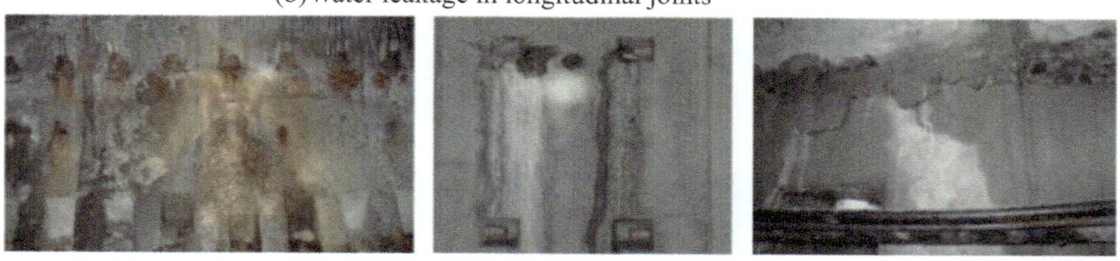

(c) Water leakage from bolt holes, cross joints of tunnel

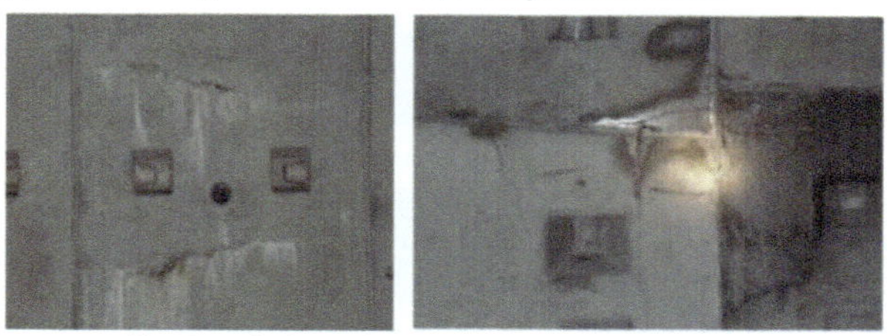

(d) Water leakage from cracks in tunnel segments

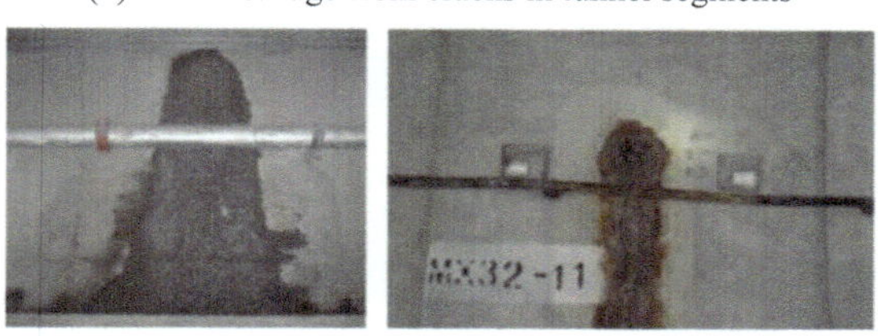

(e) Water leakage from grouting holes (or lifting holes)

Fig. 2.1 Distribution of water leakage in tunnels

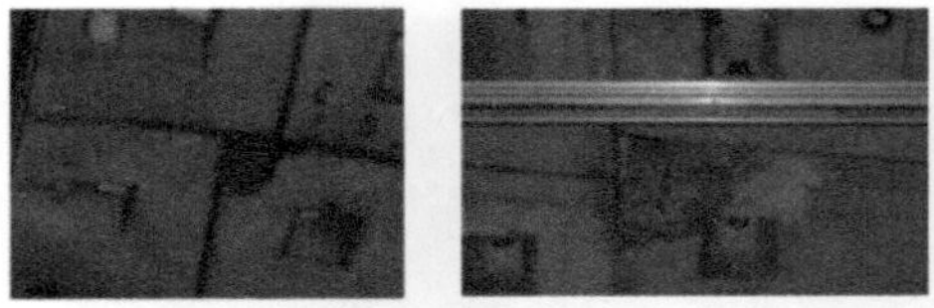

(a) Corner spalling and block spalling of segment

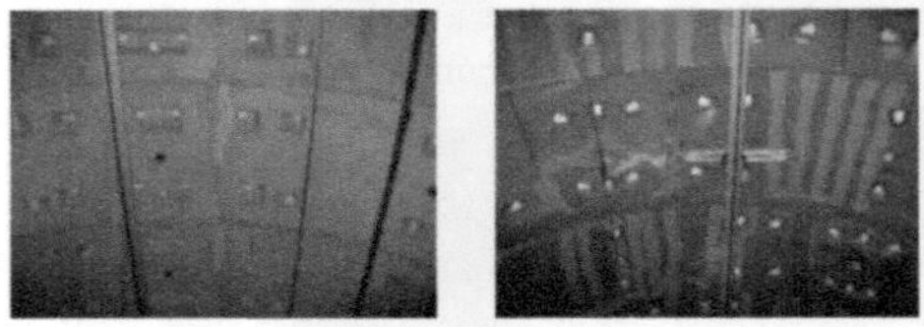

(b) Cracks in segment

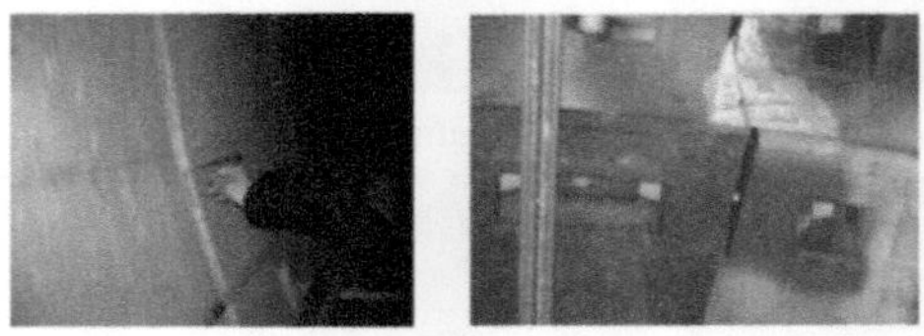

(c) Joints dislocation (d) Elongational rheological behavior of bolts

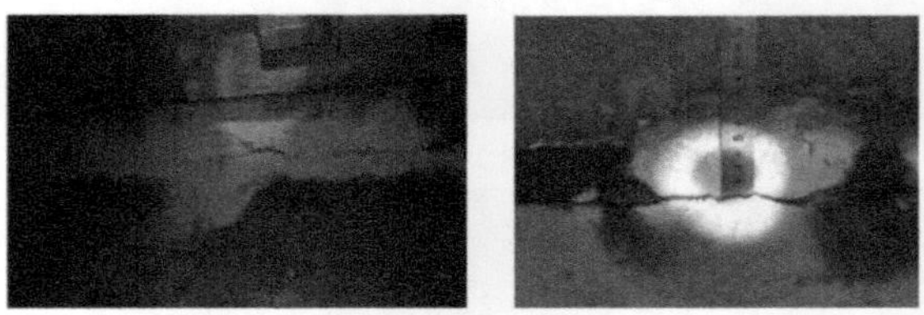

(e) Trackbed disengagement

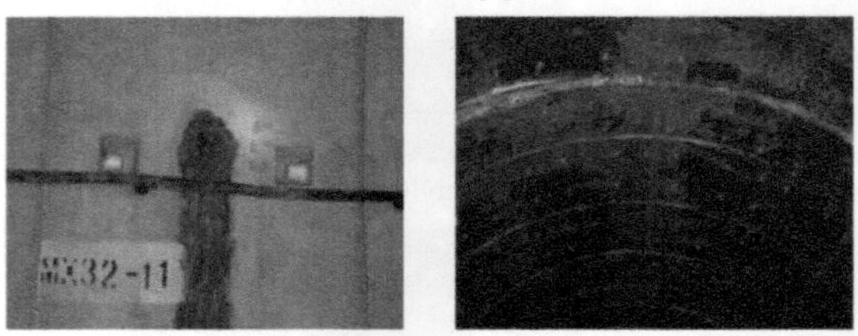

(f) Corrosion of grouting holes (g) Steel segment corrosion

(h) Springline compression damage (i) Tensile failure of bolts/elongational rheological behavior of bolts

Fig. 2.2 Damages to tunnels

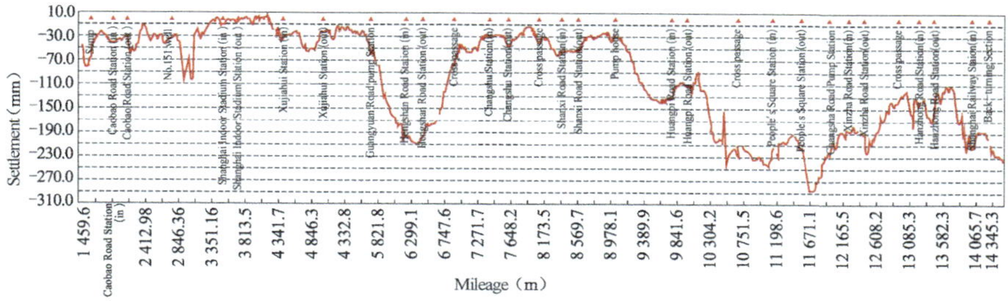

Fig. 2.3 Differential longitudinal settlement curve of Shanghai Metro Line 1

2.1.5 Transverse Convergence Deformation

Transverse convergence deformation occurs in tunnel lining in soft soil strata as the ratio of vertical and lateral tunnel loads changes due to surcharge (load from landfills, landscaping, construction of buildings, etc.), lateral unloading (water seepage and leakage, excavation of foundation pits, and other disturbances). The transverse convergence deformation will trigger new structural leakage or aggravate the structural leakage and other diseases. Such incidents have occurred in the subway tunnels of more than ten cities in China. Shown in Figs. 2.4 and 2.5 are the effect of a typical adjacent construction on the transverse convergence deformation.

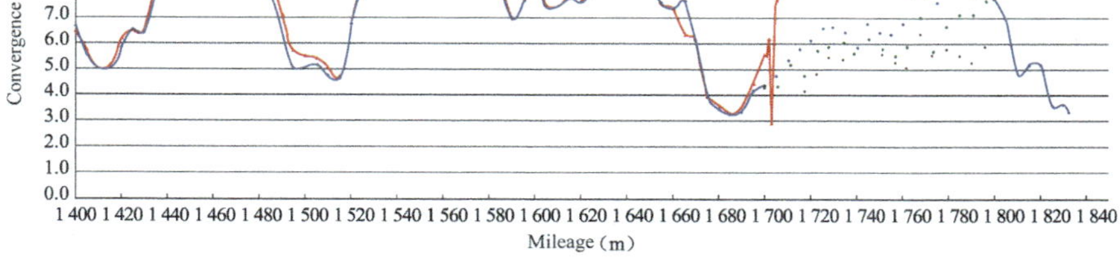

Fig. 2.4 Effect of construction on convergence deformation of Gucun Park Tunnel of Shanghai Metro Line 7

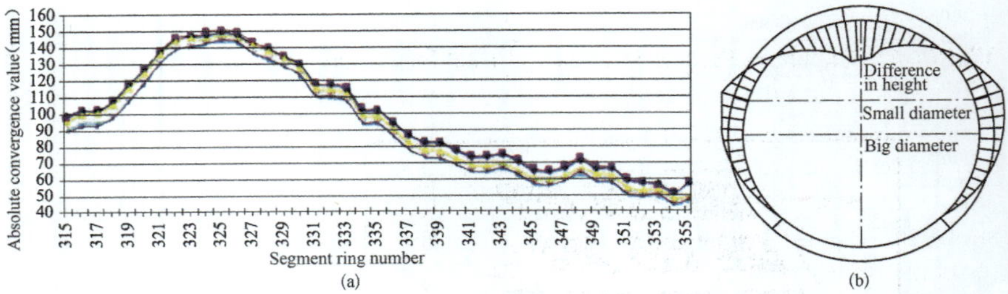

(a) The accumulated convergence curve of Xinjiangwan Block on Line 10 (resulted from lateral unload)

(b) Illustration for tunnel structural deformation

Fig. 2.5 Effect of the construction of Xinjiangwan Town Tunnel Shanghai Metro Line 10 on convergence deformation

2.2 Cause Analysis of Tunnel Structural Diseases

Many factors may trigger the tunnel structural diseases, such as inadequate consideration for tunnel planning, improper design and construction, inadequate maintenance and management during operation, and changes in the surrounding environment. All these factors can affect the tunnel structure. By years of monitoring the shield tunnels of Shanghai metro and comprehensive analysis of disease-prone sections, we have found that the occurrence and development of structural diseases in shield tunnels are closely related to the following factors.

2.2.1 Geological Conditions

There are two types of soil strata in the Shanghai area that pose a major threat to the structural safety of tunnels: soft clay stratum and sandy stratum with a high content of fine particles. These two types of soil have significant effect on tunnel structural diseases, in particular, on water seepage. Under these soil conditions, only by strict control of water seepage and leakage, maintaining the stability of the soil strata around the tunnel, and maintaining the stability of the tunnel forces, can process of tunnel deformation be kept stable. The geological conditions in Shanghai are briefly presented in the following.

1. Shallow Muddy Soft Soil Layer

The shallow stratum of Shanghai area is mainly composed of soft soil, in which exists a thick saturated high compression soft clay stratum. Being the softest stratum of soft soil, it is distributed across the downtown area of Shanghai. At present, most of the subway tunnels are built in the saturated water-bearing silt clay stratum, namely, the first soft soil stratum of Shanghai area characterized by ④ gray muddy soft clay stratum.

This kind of soil is characterized by high water content (40 ~ 60%), high porosity (1.0 ~ 1.67), medium–high sensitivity (3 ~ 7), high compressibility (compression coefficient 0.3 ~ 1.69), low density (16 ~ 18 kN/m^3), low strength (20 ~ 90 kPa), and low permeability (10^{-6} ~ 10^{-8} cm/s). The characteristics of the soil strata are shown in Table 2.2. The soil structure is prone to damage by external forces due to its medium–high compressibility and rheology. Its strength will decrease significantly because of the disturbance from the tunnel construction, resulting in large amount of consolidation deformation. Consolidation and sub-consolidation settlement occurs over a long period of time, and even lasting for quite a few years. The high compressibility, rheological properties, and thixotropy of the soil stratum will undoubtedly bring about an adverse effect on the construction and operation of the tunnels. Therefore, all the measures and methods taken must be adapted to the geological conditions when planning, designing, constructing, and operating the metro tunnels.

The distinctive features of settlement deformation in typical soft soil strata in Shanghai is characterized by "long duration and large amount". The settlement deformation lasts for a long time when the soil layer around the tunnel is disturbed. There have been some cases where tunnels built in such soils had closed for a long time for repairing due to the excessive settlement.

2. Saturated Sandy Stratum

The second type of soil stratum is the saturated sandy soil, mainly including saturated silty clay, saturated fine sand, and sandy soil. These soil strata are distributed across Shanghai. A case in point is the ②$_3$ gray silty sandy soil strata at a depth of 15 m below the ground surface of both sides of Suzhou River. This kind of loose saturated soil has high water permeability, and because the sand particles are so small that it is not easy to detect sand leakage when it occurs. There have been many instances abroad where tunnels in sandy soil layers have experienced structural failure due to the loss of fine sand. Similarly, in China, there have been multiple cases where subway tunnels in sandy soil layers have required reconstruction due to water–soil inrush induced tunnel damage. So were there cases of sand and water leakage in subway tunnels in operation service.

Fortunately, they had been found and controlled in time. In any event, water and sand leakage poses great threat to the structural safety of tunnels. In the network of Shanghai metro, most of the tunnels are built in water saturated sandy soils strata. As the sandy soil is mainly composed of fine particles of 0.1–0.005 mm in size, quicksand, often hidden, is prone to occur under the action of dynamic water pressure. When penetrative cracks appear in the tunnel lining, with the width of its cracks greater than d_{70} (i.e., the particle size of 70% of the soil particles is less than d_{70}), structural instability, or even collapse may occur due to massive water and soil erosion around the tunnel. Compared to soft clay strata, the time from sand leakage to structural damage is relatively short. The deformation, if not detected, develops

Table 2.2 Parameters of the properties of soil strata in Shanghai area

Stratum number	Name of stratum	Average depth of stratum (m)	Water content (%)	Weight capacity (kN/m^3)	Porosity e_0	Plasticity I_p	Compression modulus (MPa)	Permeability coefficient (cm/s)		Consolidation fast shear (peak)	
								Horizontal	Vertical	Cohesion c (kPa)	Angle of internal friction φ
①	Miscellaneous filling soil stratum	0 ~ 3.0									
②	Silty clay	2.0 ~ 13.0	41.0	17.9	1.16	16.5	2.75	3.2×10^{-7}	3.1×10^{-8}	23	17.0
②$_3$	Gray silty/sandy soil	2.0 ~ 15.0	39.5	18.6			8.0	3.6×10^{-5}	4.4×10^{-5}	5	30.0
③	Silt clay	7.0 ~ 10.0	47.4	17.4	1.29	17.4	3.29	3.15×10^{-6}	2.9×10^{-7}	12	20.0
④	Silty clay	12 ~ 18.0	65.0	17.0	1.44	20.1	2.26	1.47×10^{-7}	3.2×10^{-7}	11	11.5
⑤$_1$	Silty clay	22 ~ 28.0	41.7	17.6	1.1	14.5	4.8	2.98×10^{-7}	1.1×10^{-6}	17	18.5
⑤$_2$	Sandy soil	20 ~ 27.0	32.5				8.0	6.2×10^{-6}	1.3×10^{-5}	13	25.0
⑥	Silty clay	26 ~ 40.0	24.0	19.7	0.7	12.9	7.01	1.1×10^{-7}	2.8×10^{-7}	24	21.0
⑦	Sandy soil	42 ~ 52.0	27.5	19.6	0.7	10.8	11.5			3	35.0

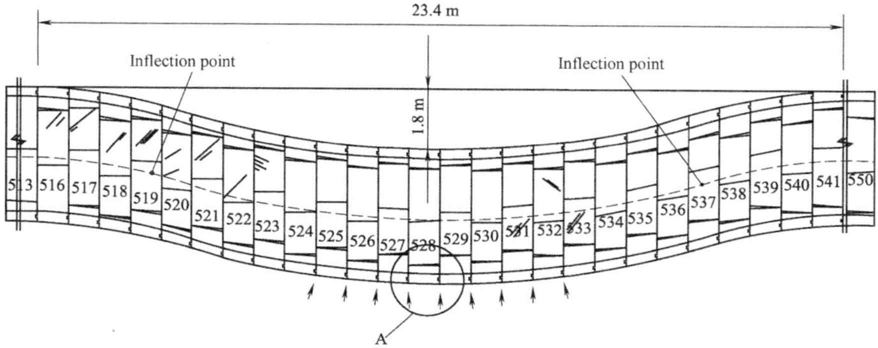

Fig. 2.6 Settlement curve of sandy soil tunnels in case of water leakage

so rapidly that the tunnel structure will soon be damaged. Figure 2.6 is a subsidence funnel formed by short-term settlement of sandy soil tunnels.

2.2.2 *Quality of Planning, Design, and Construction*

The good quality of design and construction serves as the prerequisite of the quality of the tunnel itself. With the increasing operation intensity, differential settlement and deformation are bound to increase further at the section where large deformation had occurred during the construction of station tunnels, in locations where cave in had happened, or where problems had emerged with grouting during the advancement of shield tunneling. As the settlement and deformation get worse, the tunnel fails to meet the safety standards for operation in the end. Such cases had occurred in the operating metro lines as though they differed each other in degree. As operating hours increase, improving construction had to be conducted without disrupting operations. However, the improving construction will bring about new risks to the safe operation of the metro. Therefore, it is essential that great importance be attached to settlement and deformation during the phase of tunnel planning, design, and construction to ensure the structural safety and minimize the occurrence of diseases.

1. Planning and Design Phase

Full consideration should be taken of the usability, safety, and functionality of the tunnels in operation. It is strongly recommended that the main structure of a node should be completed at one time to avoid repeated constructions at the same place. Otherwise, repeated constructions will not only affect the traffic operation but also bring greater damage to the structure, and shorten the life of the tunnel. However, in actual engineering projects, the lack of prospective planning has led to repeated adjustments in planning and design, as well as repeated construction work, resulting in many hidden hazards to structural safety. When making waterproof design for the tunnel structure, it is very important to take full consideration of and foresee the threshold of deformation and waterproof in view of the features of the soft clay in Shanghai.

2. Construction Quality of Shield Tunneling Advancement and Grouting Quality Issues

During the tunnel construction (including grouting), the excessive disturbance of the soil stratum around the tunnel will trigger significant settlement of the disturbed solid stratum, and consequently resulting in the tunnel settlement. Inert grout was widely used for grouting years before. The inert grout would segregate significantly as a result of the vibration from the passing train or over a certain period of time after grouting, bringing about adverse effect to the control of tunnel deformation. After years of research and improvement, a high-strength retardant setting grout, due to its good effect in controlling deformation and waterproof, has been widely used in the shield tunneling on the soft soil strata in Shanghai and other parts of China.

3. Disease-Prone Areas Like Portal Shafts and Pump Stations/Connected Aisles

The high risk of working in the tunnel requires that the shorter the construction time be, the better. As the construction units differ in the techniques and methods for shield tunneling, so there is rare standardized guidance for the control of construction quality. The main cause for water seepage and water leakage at the pumping stations/connected aisles is that the freezing construction is used so that problems arise in the control of the frost heave and thaw settlement to a certain degree. Therefore, it is essential that the construction pumping stations/connected aisles should follow the mandatory standards for construction. When the construction is completed, the settlement of the tunnel occurs due to the disturbance of the soil strata from all kinds of construction activities around it. The settlement and deformation of the tunnel is the result of multiple-factor interaction over a long time.

4. Subway Sections Where Dangerous Cases Had Occurred During Construction

Differential settlement and deformation aggravate further at any section where hazards had occurred during the construction of station tunnels as the operation time increases. Consequently, the tunnel fails to meet the safety standards for operation.

The dangerous cases that had emerged during the tunnel construction are shown in Fig. 2.7.

2.2.3 *Train Vibration and Vibration-Excited Subsidence*

Vibration-excited subsidence refers to the additional settlement of the soil stratum under cyclic load. The amount of soil subsidence varies in accordance with such factors as the amount of load, the stiffness of the soil body and the degree of softening of the soil body after vibration. The degree of softening of the soil body after vibration is determined by the dynamic load and the number of stress cycles. Specially, the

Fig. 2.7 Dangerous situations during tunnel construction

construction quality of the tunnel joints, the reliability of waterproof technology, the important factors as the dynamic loads on the train, the frequency of vibrations induced in the tunnel, the density of the train traffic, and the geological conditions are the main causes of the vibration-excited subsidence of the soil stratum and the tunnel. At present, the mechanical properties of the silty soil under the long-term high-density, low-frequency vibration of the train need further investigating.

Figure 2.8 shows the settlement-time curves observed between the People's Square Station and Xinzha Road Station of Shanghai Metro Line 1. It was found by monitoring device that the tunnel was basically in the state of suspension before it was put to use. And the tunnel settled at a greater rate after it was put in operation because the consolidation and settlement of the surrounding soil stratum disturbed by the advancement of the shield tunneling had not yet been fully completed, and inert grout was used for the synchronized grouting, resulting in water seepage and leakage at a few joints of the tunnel. As the traffic density increased, a large amount of settlement would occur in the soil stratum around the tunnel under the high-density and wide-frequency vibrations. In the first three years of operation, the amount of tunnel settlement could reach 2–4 cm per year, and in the following years it would become stable and the rate of settlement would decrease considerably. After five years of rapid settlement, the rate of subsidence would mitigate and become consistent with ground subsidence. Therefore, it is very important that special attention should be paid to the settlement caused by train vibration in the first few years when the tunnel was put in operation.

Both the results of indoor tests and theoretical analyses have shown that the subsidence caused by long-term vibration of trains in Shanghai metro tunnels is generally less than 5 mm given that there were no construction quality problems, no sudden changes in the soil stratum, and the disturbance of the soil stratum caused by the shield tunneling had been offset. What's more, there had been no water or mud leakage in the tunnel. Otherwise, if water leakage or mud leakage had occurred, the vibration-excited subsidence would have increased significantly. For example, the vibration-excited subsidence in the tunnels of Shanghai Metro Lines 1 and 2 in

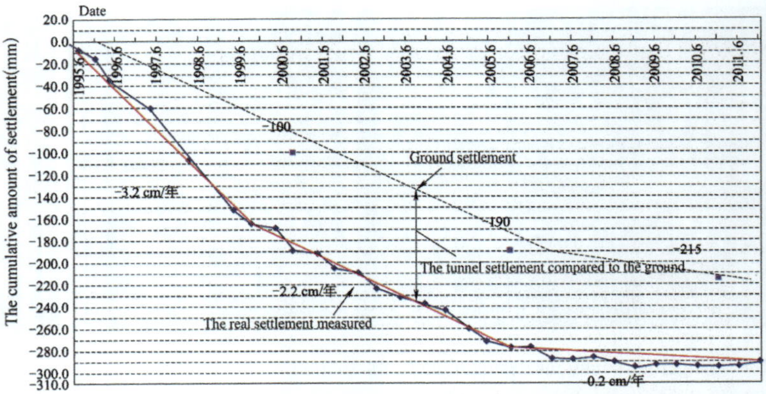

(a) Diachronic settlement curve of Line 1 Shanghai Metro at the lowest point of New World section (R713) (up)

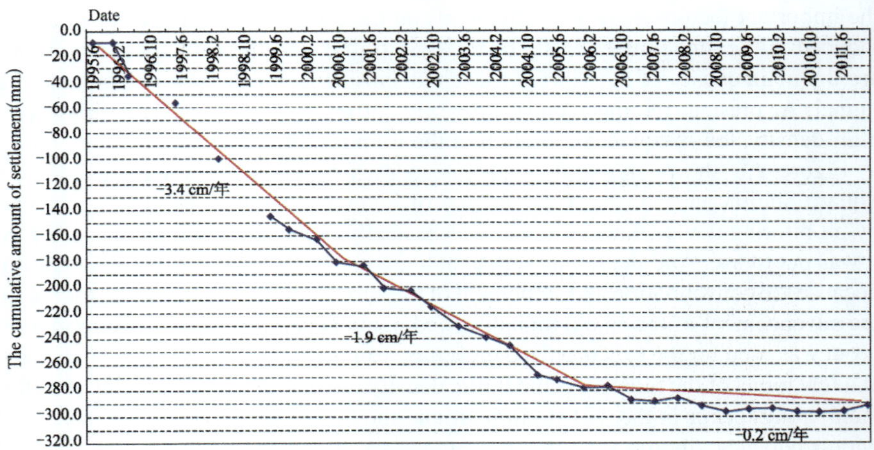

(b) Diachronic settlement curve of Line 1 Shanghai Metro at the lowest point of New World section (R713)(down)

Fig. 2.8 Settlement-time curve of Line 1 Shanghai Metro between the People's Square and Xinzha Road

quite a few sections had amounted to double digit in millimeters or even more than 100 mm in the first three years since the two lines was in operation service.

The vibration-excited subsidence is likely to emerge in the tunnel when it is put into operation if there had been disturbance or problems in synchronized grouting in the process of construction. The quality of synchronized grouting and the property of the soil stratum are the main causes for vibration-excited subsidence. The quality of synchronized grouting depends on the quality of the grout and the timeliness of grouting. Affected by train vibration, the grout would segregate due to the poor quality of grouting, leading to differential settlement of the tunnel. The underlying stratum

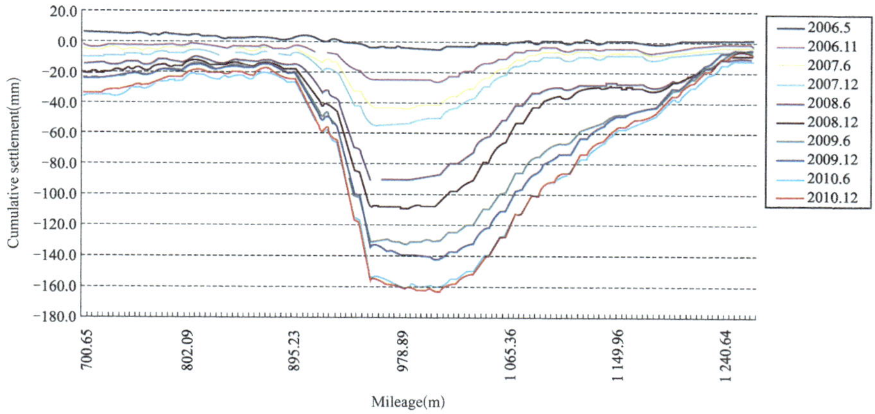

Fig. 2.9 Curve of the settlement variation detected in the subway tunnels

is prone to vibration-excited subsidence due to the long-term cyclic dynamic load of the train when the stratum is of low bearing capacity, disturbed by construction, and water seepage and leakage has occurred.

The vibration-excited subsidence trigged by long-term vibration of the train, water and soil erosion caused by water leakage interact with each other, forming a vicious circle for the structural deformation, and resulting in rapid tunnel settlement as shown in Fig. 2.9.

2.2.4 Regional Differential Settlement of Ground Surface

The overall settlement of the tunnel is not likely to affect the structural safety. However, due to numerous factors, the tunnel in fact is prone to differential settlement. The differential settlement, when exceeding the limit, will cause water leakage and even cracks in the tunnel lining, posing serious threat to the normal operation of the subway tunnel.

Regional differential ground settlement is one of the factors causing differential settlement in tunnels. The factors that cause ground settlement are many and complicated. One of the important factors is that the bedrock in Shanghai area covered with a 200–300 m mucky soil stratum has not yet completed its consolidation and settlement due to the disturbance from various activities of the city. In the past, such causes as regional deep-water extraction for production, large-scale construction activities, disturbance from ground load, and long-term consolidation of strata, had accelerated the differential ground settlement in downtown areas in Shanghai, resulting in obvious subsidence basins. The results of monitoring for the regional ground settlement at the People's Square reveal that the ground settlement around the People's Square had amounted to 1.7 m between the year of 1930 and 2000.

The monitoring results revealed that there had been more significant ground settlement in the areas or regions with higher construction intensity. The analysis of the data collected from long-term monitoring of ground settlement in Shanghai show that there is a correlation between subway settlement and ground settlement. That is, the subway tunnel that passes through the areas or regions with larger amount of ground settlement will have a larger amount of tunnel settlement accordingly, indicating that there is a good linear relationship between them to a certain extent. The ground settlement in Shanghai is not evenly distributed. The subway tunnel of Line 1 passes through a number of subsidence basins, the two subsidence basins near the Hengshan Road station and north of Huangpi Road being typical examples, as shown in Fig. 2.10. Over the past 20 years, only 1 cm or so of ground settlement had occurred near the Shanghai Stadium Station of Line 1. However, the settlement around the People's Square has reached nearly 30 cm. Statistics show that there has been a constant consistency in terms of the amount and trend of settlement from the ground monitoring points at the People's Square and the settlement detecting points in the subway tunnel since 2000. The settlements vary in other parts and in different soil strata along the subway Line 1.

Since the beginning of the twenty-first century, Shanghai has implemented strict regulations to control the deep groundwater extraction for commercial and industrial purposes, followed by laws regulating the extraction of groundwater, and stipulating that mandatory recharge measures be taken to effectively control ground subsidence. The enhancement of underground water protection has significantly controlled the settlement in the downtown area, and the subsidence rate has been reduced from the previous 10–20 mm per year to 6–7 mm per year at present.

2.2.5 *Quality of Routine Maintenance*

Structural diseases manifest as deformation, structural damage and water leakage. The quality of the routine maintenance of the tunnels is reflected by timely detection and treatment of the structural diseases. Water leakage in tunnels not only causes corrosion and damage to the tunnel facilities and equipment, but also changes the displacement and stress fields around the tunnel, leading to the change of the tunnel's stress state, and causing the instability of the tunnel structure. It is of vital importance to make timely and high-quality routine maintenance for the subway tunnels.

2.2.6 *External Environment Impact*

The external environment impact that affect the structural safety mainly includes the breakdown of the tunnel segment by unauthorized pile foundation construction, illegal load and unload, such as soil piling, excavation, prospecting and drilling, and dumping load and landscape load. Illustrated in Fig. 2.11 are some of the accident

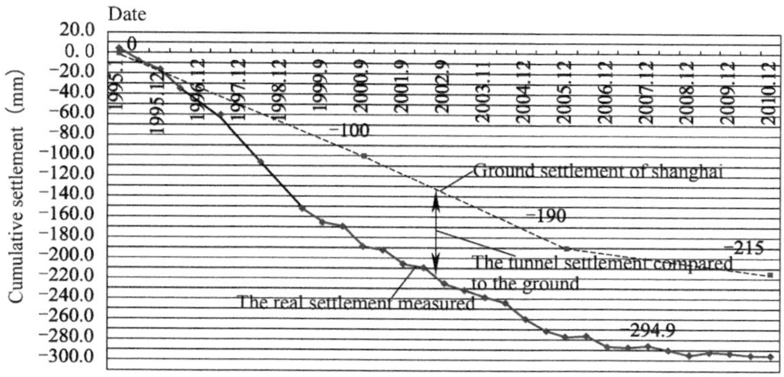

(a) Diachronic settlement curve at the lowest point of the New World (R713) Line 1 (up)

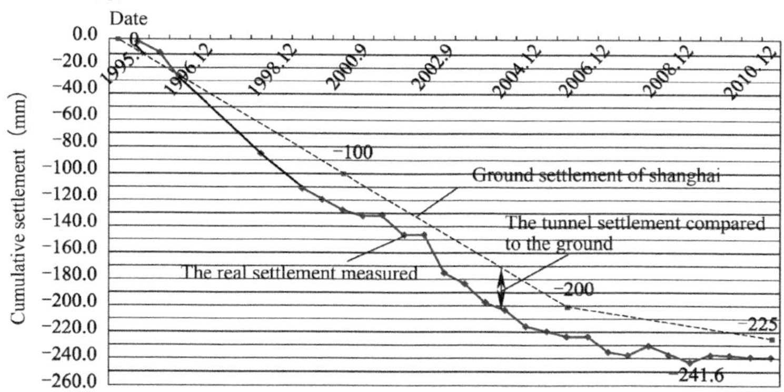

(b) Diachronic settlement curve at the lowest point of the People's Square (R684) Line 1 (up)

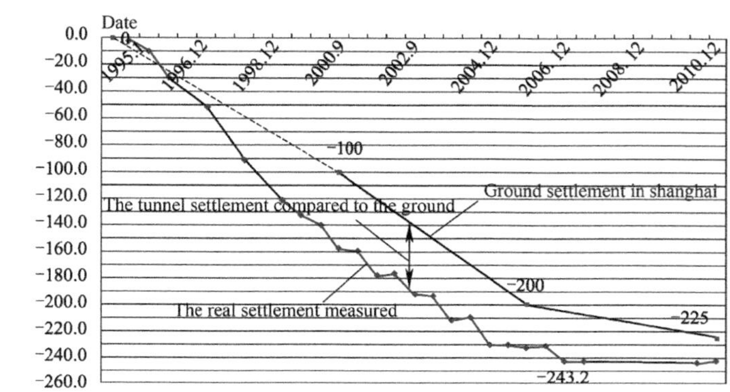

(c) Diachronic settlement curve at the pump station of Ninghai Road West (R649) Line 1 (up)

Fig. 2.10 Comparison between the settlement-time curves of typical settlement points of Line 1 and the ground settlement

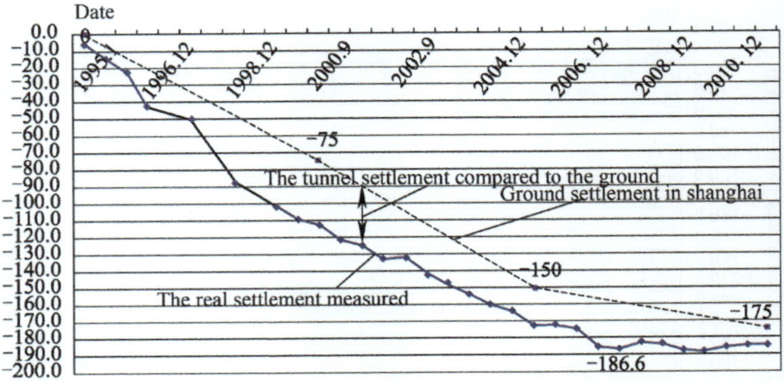

(d) Diachronic settlement curve at Hanzhong Road (R821) Line 1 (up)

Fig. 2.10 (continued)

Fig. 2.11 Destruction of the tunnel segment by external construction

sites. The external environment impact is likely to cause structural damage to tunnels, resulting in serious economic losses.

2.2.7 Impacts of Adjacent Construction Activities Along the Tunnel Route

Heave and settlement caused by the construction load or unload activities adjacent to the subway tunnel, rain water in large foundation pits in downtown areas, and soil erosion in around the tunnel caused by the construction of rain water pressure reducing facilities would have significant impact on the tunnel settlement in a large area. A case in point is the deep and large foundation pit, with tens of thousands of tons of, or as much as millions of tons of rainwater. The long period of construction would have a direct or indirect significant impact on the deformation, or even irreversible chronic impact on the tunnel structure due to the peculiar features of the soil stratum. In spite of the various control measures taken during the construction period, the disturbances would inevitably affect the tunnel structure, resulting in prolonged deformation of the tunnel.

To realize the "interconnection and one-ticket transfer" in metro network, there are naturally a large number of up and down crossings construction between the lines. In addition, a large number of major municipal underground passages are being constructed above and below the metro tunnels. It is difficult to conduct the crossing construction in soft soil because of its difficulty and relatively poor effect in control. The tunnel deformation would be exacerbated to a certain extent by the crossing construction and it could last for a long time.

The tunnel structures are frequently subject to coupling effects from adjacent construction activities along the route. The inevitable impacts cannot be ignored. The engineering practice has shown that the deformation curvature of several lines in Shanghai is approaching or exceeding the limit under the coupling effect of construction activities near the tunnels. It is necessary that much attention should be paid to impacts from the adjacent construction activities.

1. Impact of Deep Foundation Pit Excavation

The north side of the foundation pit of Jing'an Jiali Block is adjacent to the tunnel between Nanjing West Road Station and Jing'an Temple Station of Metro Line 2. The two-floor underground foundation pit has an excavated area about 11,745 m^2. The shortest distance between the pit support structure and the tunnel of Line 2 on the north side is 8 m. The excavation depth of Northern Zone I in the pit is about 14.45 m and the excavation depth in Northern Zone II and Northern Zone III is 13.95 m. The construction of the pit had triggered a serious settlement deformation in the tunnel of Metro Line 2, as shown in Fig. 2.12.

2. Impacts of Tunnel Crossing

The crossing construction of Line 11 tunnel underneath Metro Line 2, completed in June 2009, had greatly affected the settlement of the adjacent tunnel of Metro Line 2. The range of Line 11 crossing Line 2 covers K12 + 668 ~ K12 + 768. Before the end of the crossing construction, there had already been settlement of Line 2. And when the crossing construction was completed in June 2009, there had been a significant increase in the rate of subsidence, forming a large subsidence funnel. The settlement-time curves for Metro Line 2 (up) and downward are shown in Fig. 2.13. Subsequently, the settlement was brought under control by using perturbation grouting at the base of the tunnel.

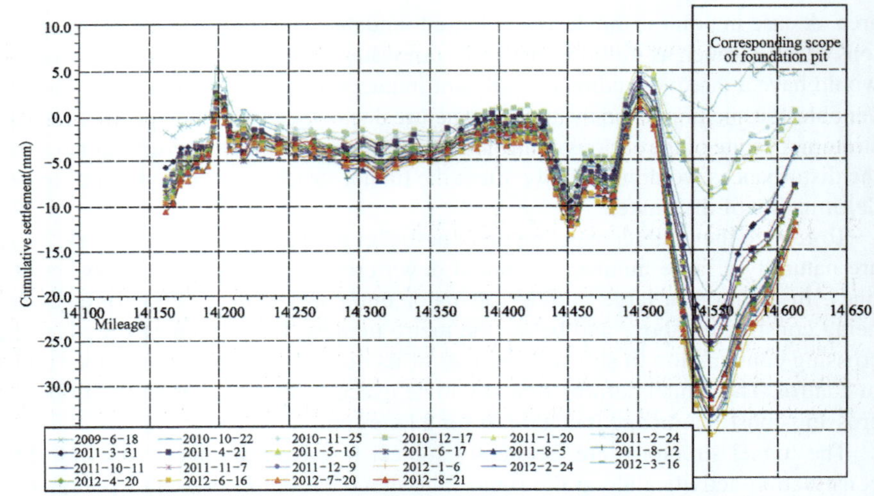

(a) Monthly cumulative Settlement at Jing'an Temple and Nanjing West Road

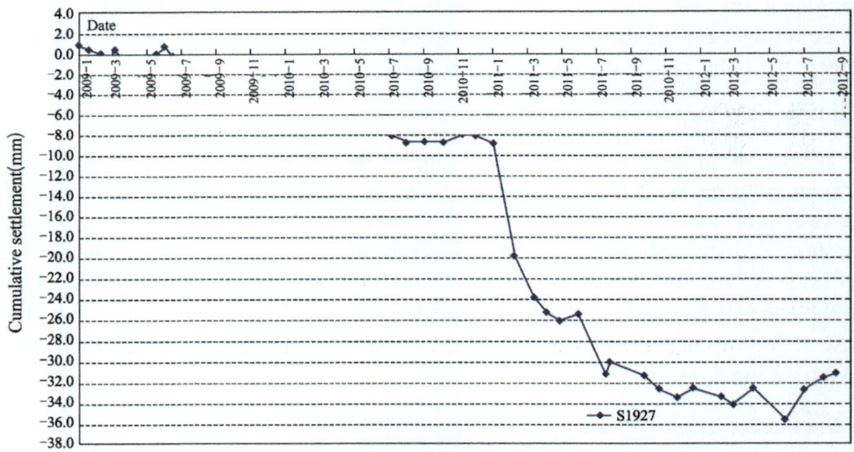

(b) Diachronic Settlement at the lowest point of Jing'an Temple and Nanjing West Road

Fig. 2.12 Cumulative settlement curve for Jing'an Temple and Nanjing West Road and settlement-time curve at the lowest point

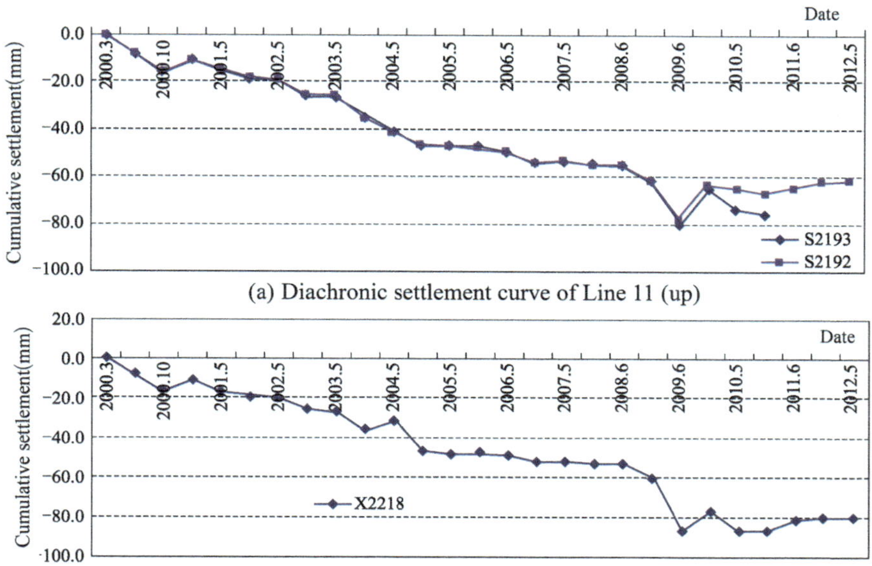

(a) Diachronic settlement curve of Line 11 (up)

(b) Diachronic settlement curve of Line 11 downward

Fig. 2.13 Settlement-time curves of Line 2 caused by Line 11 underneath crossing

2.3 Analysis of the Characteristic of Differential Longitudinal Settlement of the Tunnel

2.3.1 The Mechanism of Longitudinal Deformation

The tunnel segments are connected in longitudinal direction by a series of tenons, mortises, and bolts. The soil stratum that the tunnel passes through has different geological conditions and different rigidity in the longitudinal direction, making the longitudinal deformation of the tunnel very complicated and the deformation pattern unpredictable. According to the present research, the longitudinal deformation modes of tunnels are usually categorized into two types: rigid body rotation of the segment and joint dislocation (Wang et al. 2013).

1. Segment Rigid Body Rotation

Suppose that the ring of tunnel segment is an ideal rigid body, the deformation between the rings is mainly caused by the rotation of the rigid body in the process of longitudinal settlement deformation of the tunnel, and the rotation angle is θ. The angle of rotation is usually very small. When relative rotation of the segments occurs, the top or the bottom of the tunnel will open, the amount of openness is Δ. The deformation of the tunnel settlement is shown in Fig. 2.14. When the tunnel settles, the top of the tunnel is compressed tight, and the bottom of the tunnel opens

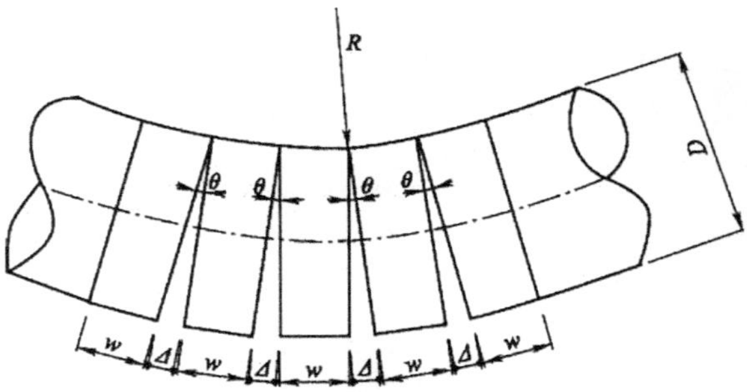

Fig. 2.14 Rigid body rotation of the whole tunnel ring

Table 2.3 Longitudinal settlement (heave) of the tunnel in relation to the rigid body rotation of the entire ring

Serial number	Radius of settlement (heave) curve R (m)	Ring joint opening Δ (mm)
1	15,000 (in accordance with protection standards)	0.413
2	10,000	0.62
3	6200	1.0
4	2066	3.0
5	1550	4.0
6	1033	6.0

(or closes). Vice versa, the top of the tunnel opens when the bottom of the tunnel is compressed tight.

According to the geometric conditions of rigid body rotation, let the width of the tunnel ring be w, the diameter D, then the geometric relationship between the opening (or closing) amount Δ and the curve radius of tunnel longitudinal settlement R is expressed as follows:

$$\frac{\Delta}{D} = \frac{w}{R} \quad \text{or} \quad \frac{\Delta}{w} = \frac{D}{R} \tag{2.3.1}$$

When the ring width takes 1.0 m and the external diameter of the tunnel is 6.2 m, the relationship between longitudinal settlement (heave) and the gap between rings is shown in Table 2.3. Based on this, the waterproofing of the tunnel would fail to function when the ring gap is up to 6 mm. But in the actual condition, the waterproofing system does not fail even when the ring gap reaches 6 mm, indicating that there is a significant discrepancy between the assumption that the longitudinal deformation of the tunnel is caused by the rigid body rotation of the whole ring and the actual longitudinal deformation of the tunnel. In the tunnels that are under construction and those that have been already built, the ratio of tunnel length to diameter L/D is usually greater than 150. As the longitudinal end of the tunnel is anchored to subway station, which has a greater rigidity, and there is a certain shear force between the tunnel and the surrounding soil as well, the longitudinal movement of the tunnel is subject to great constraint. In addition, the tightening bolts between ring segments are playing a significant role in constraining the tunnel from undergoing a rigid body rotation or large horizontal displacements (gaps) in the longitudinal direction. Generally, under the circumstances, large inter-ring gaps or rigid body rotations are not likely to occur in the longitudinal direction of the tunnel.

2. Joint Dislocation

Joint dislocation is likely to occur in the process of longitudinal settlement deformation of the tunnel due to connection methods of the rings and the lower shearing force. For example, suppose that the settlement curve of the tunnel be circular, let the ring at the lowest point be the first ring, namely, the datum point. The difference between the bottom of the tunnel of the first ring and the lowest point of the settlement—curve is defined as the initial dislocation δ_1. The dislocation deformation between Ring 2 and Ring 1 is defined as δ_2, and the joint dislocation between ring i and ring $i-1$ is defined as δ_i. According to Fig. 2.19, the amount of dislocation δ_i of ring i is calculated with equation

$$\delta_i = \sqrt{R^2 - (i-1)^2 w^2} - \sqrt{R^2 - i^2 w^2} \tag{2.3.2}$$

where w is width of the ring.

The cumulative longitudinal deformation is calculated by

$$\sum \delta_i = R - \sqrt{R^2 - i^2 w^2} \tag{2.3.3}$$

Settlement Radius $R = 15{,}000 \sim 100$ m. The variation of the dislocation between segment rings is shown in Table 2.4, and the settlement curves formed by dislocation between different segment rings are shown in Fig. 2.15.

Analysis results of Table 2.4 and Fig. 2.15 show that:

(1) The larger radius of the settlement curve, and the larger range of settlement, the smaller joint dislocation between rings; vice versa, the smaller radius of the settlement curve, the smaller range of settlement, and the larger joint dislocation between rings (i.e., the joint dislocation will soon exceed the threshold).
(2) The larger radius of the settlement curve, the greater cumulative settlement in the settlement area. Even if there is a small amount of joint dislocation between rings, the cumulative deformation in a wide area is still very significant.
(3) On the same settlement section, the further away from the datum point, the greater joint dislocation deformation between rings.

The safety of the tunnel depends on the safety of the tunnel structure and the waterproofing system. The long-term on-site supervision and monitoring of tunnels has found that the settlement and deformation of tunnels and water leakage act upon each other. Water leakage will aggravate the deformation of the tunnel. The

Table 2.4 Variation of joint dislocation with the same settlement radius

Joint dislocation (mm)	R (m)						
	15,000	10,000	3000	2000	1000	500	100
Initial value δ_1	0.0333	0.05	0.1666	0.25	0.5	1.0	5.0
Ring 2 δ_2	0.1	0.15	0.5	0.75	1.5	3.0	15.0
Ring 3 δ_3	0.1666	0.25	0.8333	1.25	2.5	5	25.008
Ring 4 δ_4	0.2333	0.35	1.1666	1.75	3.5	7.0	35.022
Ring 5 δ_5	0.3	0.45	1.5	2.25	4.5	9.0	45.046
Ring 6 δ_6	0.3666	0.55	1.8333	2.75	5.5	11.0	55.084
Ring 7 δ_7	0.4333	0.65	2.1666	3.25	6.5	13.0	65.139
Ring 8 δ_8	0.5	0.75	2.5	3.75	7.5	15.0	75.213
Ring 9 δ_9	0.5666	0.85	2.8333	4.25	8.5	17.0	85.31
Ring 10 δ_{10}	0.6333	0.95	3.1666	4.75	9.5	19.0	95.433
Ring 11 δ_{11}	0.7	1.05	3.5	5.25	10.5	21.0	105.585
Ring 12 δ_{12}	0.7666	1.15	3.8333	5.75	11.5	23.006	115.77
Ring 13 δ_{13}	0.8333	1.25	4.1666	6.25	12.5	25.0078	125.99
Ring 14 δ_{14}	0.9	1.35	4.5	6.75	13.5012	27.01	136.249
Ring 15 δ_{15}	0.96666	1.45	4.8333	7.25	14.5015	29.012	146.55
Ring 16 δ_{16}	1.0333	1.55	5.1667	7.75	15.5018	31.015	156.898
Ring 17 δ_{17}	1.1	1.65	5.5	8.25	16.5	33.018	167.295
Ring 18 δ_{18}	1.1666	1.75	5.8334	8.75	17.5027	35.0215	177.745
Ring 19 δ_{19}	1.2333	1.85	6.1668	9.25	18.503	37.0254	188.252
Ring 20 δ_{20}	1.3	1.95	6.5	9.75	19.5037	39.03	198.819

Note If the calculated value in the table exceeds 8 mm, it is set as the threshold for joint dislocation control

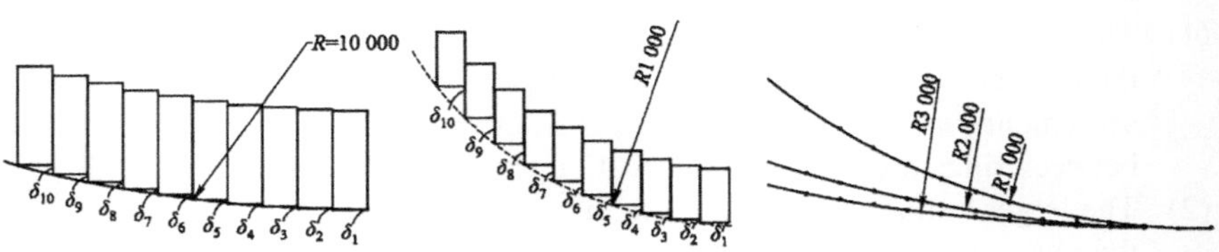

Fig. 2.15 Joint dislocation at the same location with different settlement radii

worsened deformation will in turn aggravate water leakage in the tunnel, forming a vicious cycle.

By analyzing the data from monitoring longitudinal deformation of the tunnels operating in Shanghai, it has been found that the deformation of the rigid body rotation of the segment ring is not likely to occur. As shown in Fig. 2.16a, no such deformation

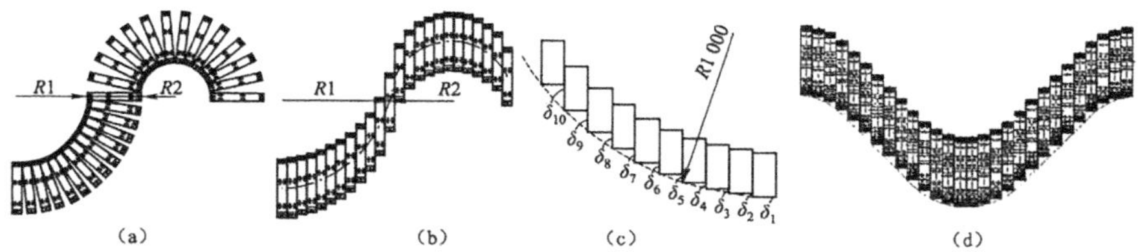

Fig. 2.16 Characteristics of the longitudinal deformation of the tunnel

patterns are found in either the tunnels under construction or in operation. The reason is that the two ends of a section of tunnel are anchored to the shaft or station opening that have already been fixed. The displacement of the tunnel end is constrained by the fixed joints and the surrounding soil. The tunnel cannot extend in horizontal direction, and it is not likely to bend as to trigger deformation. Therefore, the longitudinal deformation of the tunnel can only take place in the form of joint dislocation between the rings, forming a joint dislocation settlement curve. The longitudinal deformation often occurs between the joints of two rings in the form of joint dislocation. The deformation curve is shown in Fig. 2.16b–d. The settlement of the tunnel can be a single curve or multiple curves. It can be deduced by the characteristics of the curves that the track bed are disengaged from the segment rings as it had been observed in the site inspection.

Theoretical analyses and field measurements show that the longitudinal deformation of the tunnel is in the form of joint dislocation. It is rational to assume that the longitudinal settlement take the form of joint dislocation between a series of segment rings.

3. Analysis of the Impact of Longitudinal Settlement in Tunnels

The total length of the tunnel increases when the settlement (heave) occurs. The more settlement variations, the greater amount of settlement, the greater increase in the *total length of the tunnel. When the amount of joint dislocation is small, the amount of* longitudinal deformation increased is accordingly small, which is calculated as follows:

$$\Delta l = \sqrt{(w + W_0) + \delta_i^2} - w \tag{2.3.4}$$

where

Δl is the increased longitudinal deformation of the tunnel.
w is the tunnel settlement.
W_0 is the added longitudinal settlement between segment rings gaps.
δ_i is the amount of joint dislocation of Ring No i.

The joint dislocation deformation is shown in Fig. 2.17.

When the joint dislocation exceeds the range of 4–8 mm, we must take into account the increased longitudinal ring gap W_0 when calculating of the longitudinal length

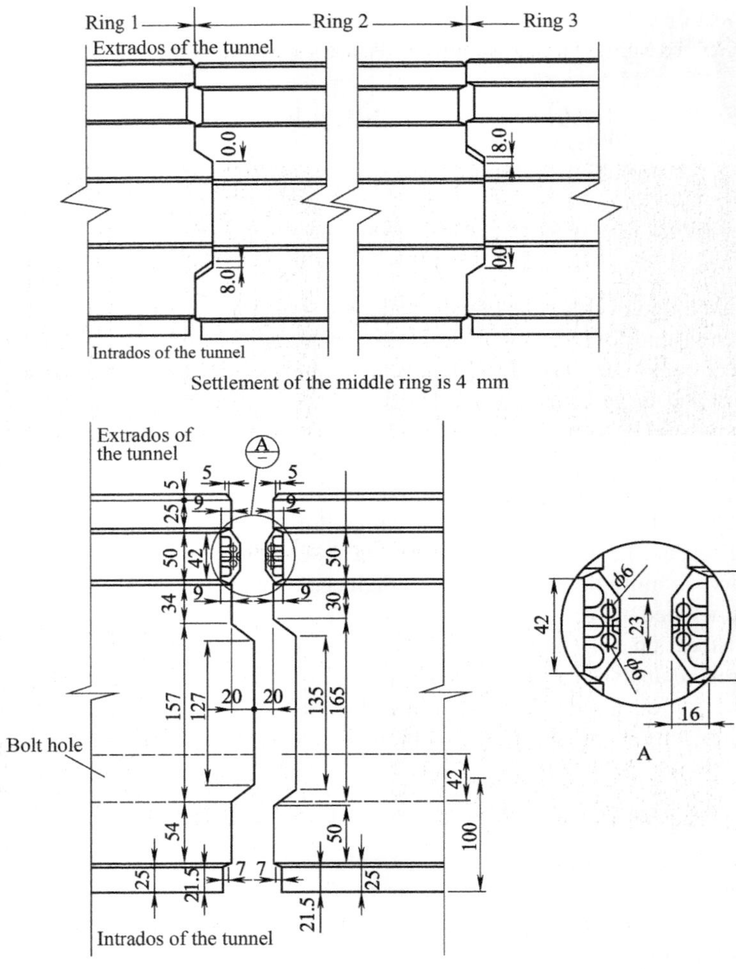

Fig. 2.17 Enlarged view of the joint dislocation of the 3-ring tunnel and the waterproof rubber strip (mm)

of the tunnel. The impact of on the structural safety and waterproofing of tunnels can be analyzed according to the joint dislocation in different degrees.

(1) When the joint dislocation is 1 ~ 4 mm. The joint dislocation can be regulated by the design of the tunnel ring configuration itself, but it also creates a certain amount of tension and compression on the gasket. Geometrically, the radial compression of the gasket before deformation can be at least approx. 23 mm, and it still remains approx. 19 mm after dislocation. According to Eq. (2.3.4), if the joint dislocation is 1 mm, the length of single-ring tunnel will increase in

0.005 mm. If the joint dislocation between rings is 4 mm, the length of single-ring tunnel will increase 0.008 mm. When the joint location is small, the increase in longitudinal length of the tunnel caused by it will accordingly be small. The gap between segment rings does not increase either. As the joint dislocation increases, different parts of the gasket are in complex force states, such as tension, compression, shear, and so on. Although the gasket does not leak, the waterproofing ability between the rings is significantly weakened increasing greatly the probability of water leakage in the tunnel. The longitudinal coupling bolts may further exerts a tensile effect and a compressive shear effect on the concrete in the hand-hole area.

(2) When the joint dislocation between rings reaches 4 ~ 8 mm, tenon and mortise groove is in the ultimate state of deformation. At this moment, the top edge of the tenon contacts the bottom edge of the mortise, exerting a direct shearing force. The bolts are also in the state of further tension. The deformation of the sealing gasket and the state of the force aggravate with the increase of joint dislocation, the radial compression of the sealing gasket can still be up to 15 mm. According to Eq. (2.3.4), if the joint dislocation reaches 4 ~ 8 mm, the length of single-ring tunnel will increase 0.032 mm. Under this condition, the increase in the total length of the tunnel caused by joint dislocation is still relatively small. Though the gap between the segment rings remains unchanged, still the waterproofing capacity between the rings is severely weakened, and the chances of water leakage in the tunnel will be multiplied. Water leakage, once it occurs, will deteriorate the joint dislocation of the segment rings.

(3) When the joint dislocation between rings reaches 8 ~ 13 mm. Further shearing of the tenon and mortise groove on circumferential joint will cause local cracks in the grooves, leading to waterproofing failure. At this point the tenon and mortise groove on circumferential joint will slide along the contact surface, the gap between the segment rings expands linearly, leading to elongational rheological behavior of bolts. At this time, although the radial compression of the gasket still remains 10–15 mm, the circumferential gap between the rings has exceeded the waterproof threshold due to local damage to the segments. According to Eq. (2.3.4), if the joint dislocation between the rings reaches 13 mm, the length of the tunnel increases rapidly. It can reach 13.083 mm for a single ring, and the gap between the rings increases rapidly by more than 6 mm, the waterproofing system between the rings basically fails to function, leading to serious water seepage. When vertical joint dislocation occurs, the dislocation of the gasket attached to the segment ring occurs accordingly as shown in Fig. 2.18.

The joint dislocation occurring at the top or bottom of the tunnel are obvious. The tenon and mortise groove on circumferential joint are shearing and sliding, the joint dislocation develops linearly. The tunnel is in critical condition at this point. The risk is that if the deformation continues to develop, it will seriously affect the structural safety of the tunnel.

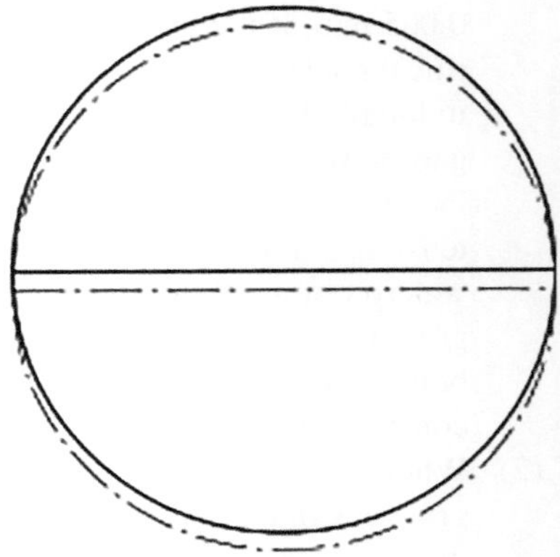

Fig. 2.18 Joint dislocation of the whole segment ring

(4) When the joint dislocation fall between 13 and 23 mm, the constant shearing will cause further damage to the tenon and mortise structure, resulting in total failure in waterproofing. Though the structure of tenon and mortise are still in contact, the joint dislocation will develop rapidly in a linear way, leading to the instability of the tunnel structure. If the underlying soil stratum is sandy soil, the structural safety of the tunnel is at even greater risk.

To sum up, when the joint dislocation is relatively small (generally less than 10 mm), the increase in the total length of the tunnel is small, and the gap between the segment rings does not increase. Though the structural safety of the tunnel is still controllable, the waterproof effect of the sealing gasket will be greatly weakened. If the joint dislocation reaches its ultimate limit, the gap between the segment rings will develop rapidly in a linear way, resulting in cracks in segments and failure in waterproofing. This will bring catastrophic threat to the safety of the tunnel. Therefore, the increase in joint dislocation will not only cause shearing of the tunnel ring, but also lead to an increase in the longitudinal displacement (ring gaps) of the tunnel.

The above study only made analysis of joint dislocation of the tunnel. In fact, since the longitudinal deformation is complex, and the transverse deformation will occur at the same time after the tunnel construction is completed, the installation relation between the tenon and mortise also varies. Due to the interaction between the tunnel segments, the longitudinal and transverse deformations of the tunnel interact with each other to a certain extent, making the deformations of the tunnel even more complex and unpredictable.

According to the principle "three points fix a circle", the radius R of the additional settlement curve formed by the joint dislocation of the segment rings is shown in Fig. 2.19, the value of which is calculated as:

$$R = \frac{L^2}{2\delta} \tag{2.3.5}$$

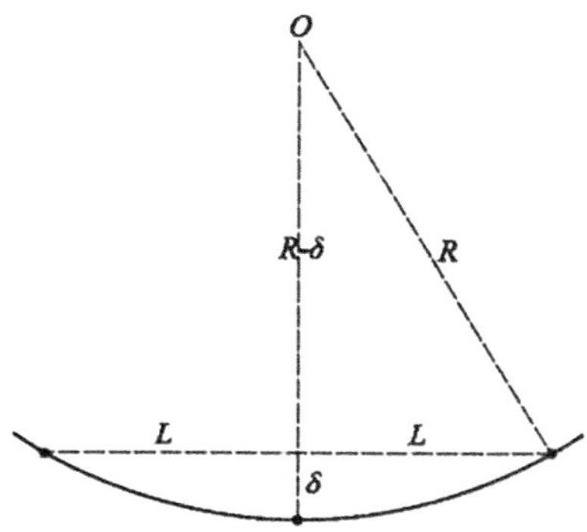

Fig. 2.19 Relationship between unattached longitudinal settlement curves for tunnels

Table 2.5 Relationship between differential longitudinal deformation and additional settlement radius of the tunnel

Serial number	Additional settlement (heave) curve radius (m)	Differential longitudinal settlement of the tunnel δ (mm)
1	15,000 (in accordance with protection standard)	0.033
2	1000	0.5
3	500	1.0
4	100	5.0 (waterproofing failure)
5	66.7	8.0 (the gap of tenon and mortise in vertical fit is in its ultimate limit)
6	50	10.0 (waterproofing may fail)

where

L refers to 1/2 of the length of the longitudinal settlement groove occurring in the tunnel.
R refers to radius of the additional curve of settlement.
δ refers to the difference in settlement within the vertical settlement limit.

Taking as an example, the ring width of 1 m, and the external diameter of the tunnel of 6.2 m, to satisfy the deformation protection standard of the tunnel, the additional differential settlement between the two rings should be strictly controlled at 0.0333 mm or less. Therefore, the joint dislocation between the two segment rings should be controlled at 0.0333 mm or less to meet the tunnel protection standard. The relationship between the longitudinal deformation of the tunnel and the additional settlement curve for a tunnel with a ring width of 1 m and external diameter of 6.2 m is shown in Table 2.5. It shows that the tunnel has the ability to adapt to the small radius of the additional settlement.

2.3.2 Case Study

1. Beizhai Road Parking Line

Figure 2.20 shows the settlement curve of a train standing line downward at Beizhai Road Station Shanghai Metro Line 2 after an accident at its pumping station. The construction of the pumping station triggered a large amount of water and sand leakage, resulting in severe deformation. The joint dislocation between some rings reached several centimeters. The maximum cumulative settlement amounted to 26 cm. The risk was brought under control after emergency repair work was conducted in time.

The cumulative settlement was up to 26 cm. The maximum joint dislocation reached 4.3 cm at two locations. The joint dislocation between the rings reached 2 ~ 3 cm for more than 10 segment rings, causing water and sand leakage between the rings of the tunnel. Although the risk was controlled by increasing the internal bracing and grouting, significant settlement occurred inside the tunnel and on the ground. The radius of the smallest additional settlement curve was less than 80 m. The tunnel structure lost its support because of soil and sand leakage. The more water and soil from the underlying soil stratum to the tunnel will result in the more serious longitudinal deformation of the tunnel, which will cause greater opening of the circumferential joints at the bottom of the tunnel. The interaction between the water and soil erosion and increased gap of the segment joints will ultimately result in damage to the tunnel.

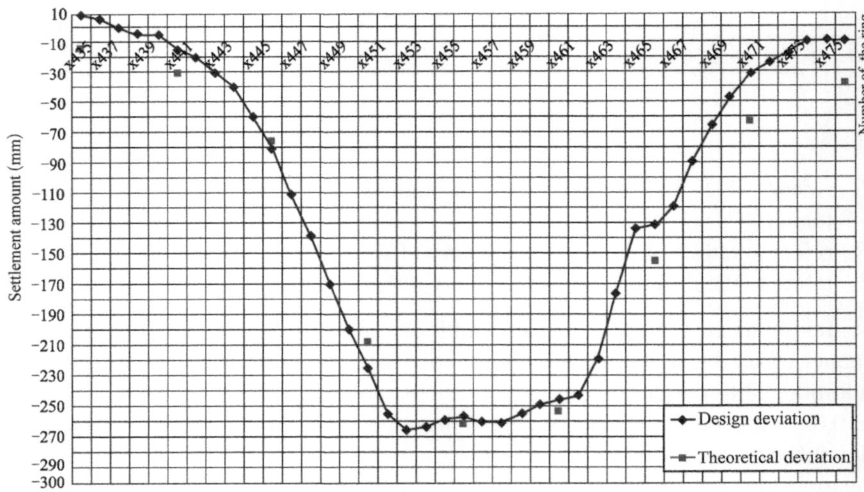

Fig. 2.20 Settlement curves formed after an accident in a standing line tunnel

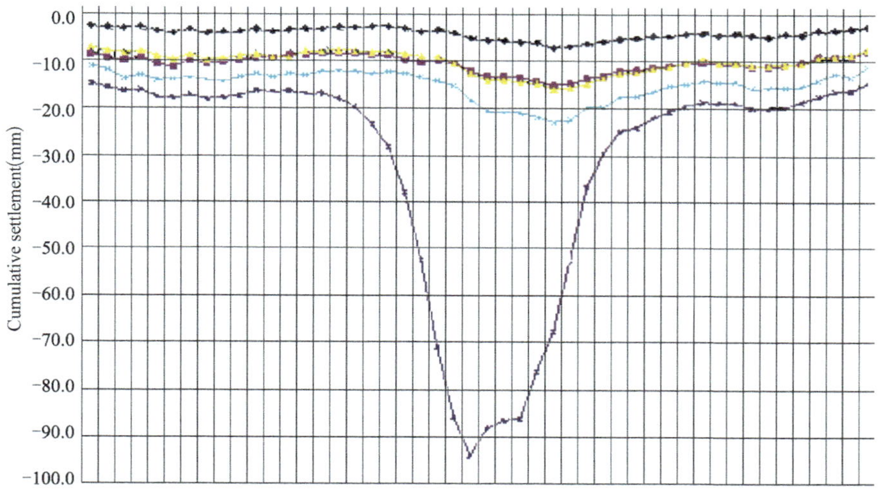

Fig. 2.21 Settlement curve formed after sudden sand and water leakage in running tunnel (down) of Line 4

2. Line 4 Dalian Road Running Tunnel

Figure 2.21 is the settlement curve formed after the sudden sand and water leakage in the tunnel of the Dalian Road running tunnel (down) of Line 4. The inherent defects of the tunnel structure are the main cause of water and sand leakage in the tunnel, triggering a large joint dislocation. The amount of joint dislocation reached 3 ~ 5 mm, and the cumulative settlement was up to 90 mm, affecting an area of more than 100 m in distance. Emergency repair work was done in time and the deformation was perfectly treated.

3. Line 18 Yingchun Road Running Tunnel

Figure 2.22 shows the settlement curve formed after the sudden sand and water leakage in the running tunnel of the Yingchun Road Line 18. The freezing construction triggered water and sand leakage in the tunnel. In consequence, a large joint dislocation occurred, reaching 100 mm, and the cumulative settlement reaching 50 cm. Areas within 20 m in distance were apparently affected. Fortunately, it was found in time and emergency repair work was carried out to eradicate the risk.

Excessive deformations of the tunnel structure due to large joint dislocation will inevitably lead to cracking of the tunnel structure, resulting in damage or destruction of the tunnel and failure of the waterproofing system, posing a direct threat to the tunnel structural safety, as proven by the longitudinal deformations that have occurred in many tunnels.

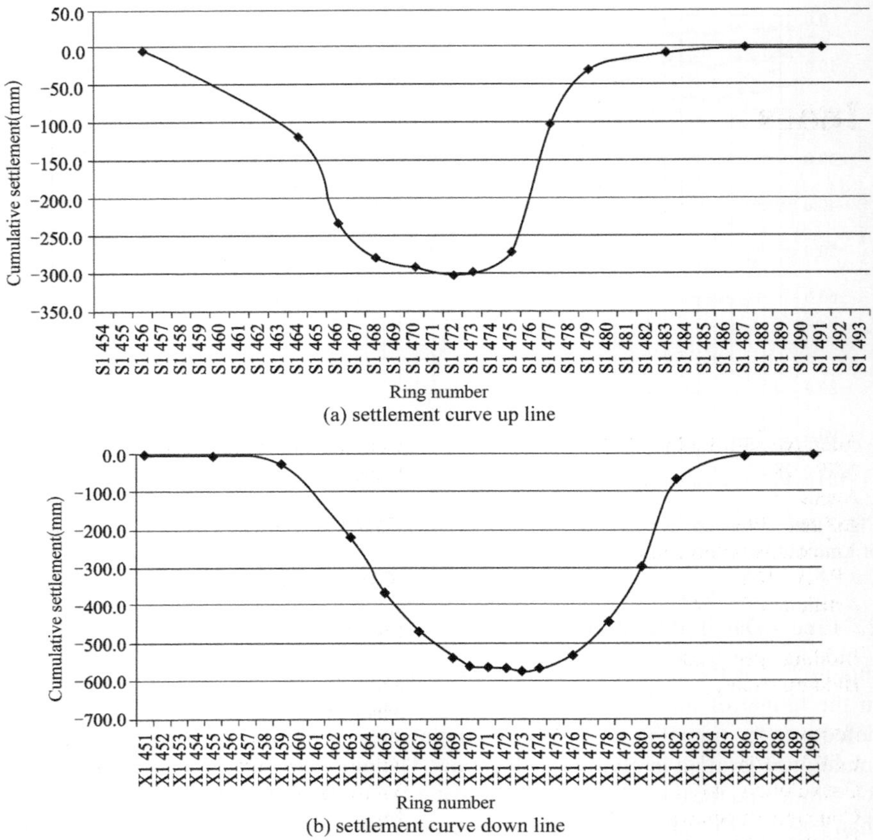

Fig. 2.22 Settlement after sudden water and sand leakage in the Yingchun Road running tunnel of Line 18

2.3.3 Control of the Differential Longitudinal Settlement in Tunnel

Based on the theoretical research of differential longitudinal settlement of the tunnel triggered by joint dislocation, and by analyzing the data collected from long-term monitoring the following deformation control indicators are suggested:

(1) Radius of curvature R being not less than 3000 m. To keep the smoothness of an operating train, the track height tolerance should be less than 4 mm/10 m, so the corresponding radius of curvature should be 3000 m. If the rails and track bed are fastened tight to the tunnel segments, the radius of curvature should take 3000 m as the threshold for the control of differential longitudinal settlement in tunnel.

(2) Differential settlement gradient being no larger than 0.16%. For continuous and extensive differential settlement, the gradient of differential settlement can be used as an indicator for assessing structural deformation.
(3) The differential settlement rate being less than 0.6 mm/month in a year. If differential settlement aggravates to the extent that the radius of curvature of the cumulative settlement curve exceeds the threshold or the gradient of differential settlement exceeds the indicator, measures should be taken to control the settlement.
(4) In the first three years when the tunnel is in operation, the settlement deformation is generally large because of the soil disturbance and consolidation after the advancement of the shield and the dynamic load of the train. Construction practice has found when the settlement is greater than 2 cm/year, and the settlement is accelerating in the tunnel. It is necessary to take special measures to control the settlement.

2.4 Lateral Convergence Deformation of Tunnel Lining Ring

2.4.1 Mechanism of Lateral Deformation

Many factors may cause lateral deformation in tunnels, including those from tunnel installation, or the comprehensive impact of multiple factors after the years' operation of the tunnel. The results of long-term monitoring revealed that two main causes that may trigger large lateral deformations in the operating tunnels. The first one is the upper load of the tunnel, and the second one is the unload disturbances from the excavation on both sides of the tunnels. If these two conditions occur at the same time, the ratio of the upper loads to the horizontal resistances and lateral stresses increases dramatically, which has an even more significant impact on the deformations of the tunnels.

1. Upper Load of the Tunnel

When the upper load of the tunnel is small, it will cause an increase in the internal force of the tunnel structure and an increase in the lateral deformation of the tunnel. At the same time, it will compress the lateral soil on the outer wall of the tunnel, increase the lateral resistance of the tunnel, and prevent or slow down the lateral deformation of the tunnel. However, it also results in the generation of excess pore water pressure around the tunnel at the same time. When the excess pore water pressure dissipates, the lateral deformation to continue to increase, and the tunnel deformation gradually reaches equilibrium.

As the upper load increases, the internal forces in the tunnel structure will increase, and the lateral deformation increases, as a result, the lateral resistance of the tunnel increases further. If the internal forces in the tunnel structure are still less than the permissible stresses after load, the structure remains stable; if the internal forces

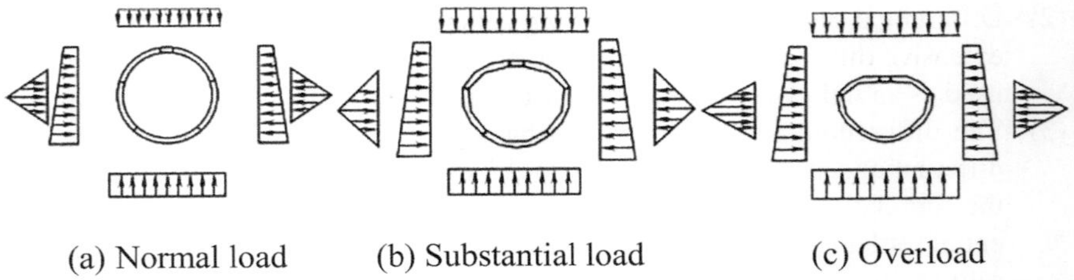

Fig. 2.23 Schematic diagram of deformation due to the variations in upper load and lateral resistance of the tunnel

exceed the permissible stresses, the segment starts to crush or crack and the bolts will be ineffective due to its elongational and rheological behavior. The tunnel is taken as an assembled integral structure, and when the structure is stressed to its limit somewhere, the stresses will be readjusted within the tunnel structure. With the gradual increase of the loads on the tunnel, deformation and stresses on the segment expand in a progressive manner until the overall structure breaks down. The load and tunnel deformation are shown in Fig. 2.23.

2. Lateral Unload

When the load at the top of the tunnel remains unchanged, but the unload construction on both sides of the tunnel will disturb the lateral soil, cause a significant loss of lateral pressure and resistance, which will rapidly increase the lateral deformation of the tunnel. If the lateral pressure loss is too much, the joints between the segment rings will open or close under pressure, resulting in the localized crushing or cracking of the concrete, and elongational and rheological behavior of the bolts. This will quickly jeopardize the overall stability of the structure and cause the tunnel collapse.

3. Upper Load Coupling Lateral Unload

The most unfavorable conditions for tunnel deformation and forces are those in which the tunnel is loaded on the top and laterally unloaded at the same time. The superposition of the operation conditions will accelerate the development of tunnel deformation and stresses, causing damages to the tunnel.

In terms of the nature of the lateral deformation of the tunnel structure, the increase in the ratio of the upper load to the horizontal load (including lateral pressures and resistances of the soil) is the core that causes the increase in lateral deformation of the tunnel. Therefore, it can be said that crush and damage of the segments, the loss of horizontal resistance are main causes of the collapse of a tunnel. If the ratio of vertical load to horizontal force acting on the upper part of the tunnel is small, the tunnel structure will be stable and less likely to collapse. If the ratio is too large, the tunnel will collapse even if the upper load of the tunnel is relatively small. But the assumption of this mode of damage and the magnitude of the transverse deformation in the event of damage need verifying by tests.

2.4.2 Characteristics of Lateral Convergence Deformation of Tunnel

Since the tunnels are installed from segments, the circumferential rigidity along the lining is different. The sum of the deformation of the installation and the deformation under stress must not exceed 5‰D according to the design requirements (Ministry of Housing and Urban-Rural Development of China 2008) (D refers to the external diameter of the tunnel), with the gap of joints no more than 4 mm. However, the actual monitoring found that when the gap of joints is more than 4 mm, the diameter change ΔD generally reaches or exceeds 5‰D. But in fact chances are that the joint deformation is often larger than 5‰D. The past data show that the tunnels of lateral deformation subject to vertical compression often take the shape of a lying "duck egg", while only a very few of tunnels look like a standing "duck egg" because of certain special causes. When the lining ring structure is stressed, the top and bottom inner sides are under tension and outer sides are under compression; the inner sides of the left and right side are under compression and outer sides are under tension. In the case of straight-joint and staggered-joint tunnels, due to the difference in stiffness of the joints, the straight-joint convergence deformation occurs along the joints, while the staggered-joint deformation occurs due to segment cracking. The measure results also revealed that the convergence deformation curves of the tunnel were not very smooth, and the transverse deformations at the top and bottom of the joints showed some abrupt variation. The main causes are the relatively low stiffness of the joints and the large joint opening. It is concluded according to the comprehensive analysis that the transverse deformation of the tunnel is mainly composed of two types: the joint deformation and the elastic deformation of the segments, of which the former is dominant, as shown in Fig. 2.24.

In general, when the extrados at the top and the bottom of the tunnel is under compression, the intrados at the top and the bottom is in tension, and external side of the springline is under tension and the inner side under compression. When the extrados at the top of the tunnel is under compression, and the extrados of the joint

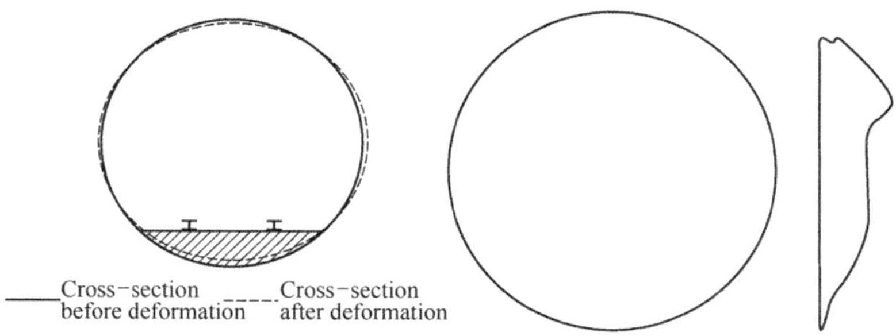

(a) Transverse deformation of tunnel (b) Distribution of transverse deformation of tunnel

Fig. 2.24 Characteristics of convergence deformation of the tunnel

is compressed, water leakage is not likely to occur. But when the compressive stress exceeds the concrete strength of the joint, damage will occur at the contact of the segment. The spring between the longitudinal joints can, to some extent, serve to dissipate the compressive stresses on the joints near the extrados.

Assuming that the initial state between the segments is tightly compressed, according to the position of the gasket and bolts, when the outer arc of the joint is pressed while the inner arc is opened by 1 mm, the length of the bolts will increase about 0.657 mm, and the opening of the gasket will increase 0.157 mm. According to the design requirements, when the threshold for control joint opening is 4 mm, the bolts will elongate 2.628 mm, and the elastic gasket opening will increase 0.628 mm. Therefore, when the outer arc of the joint is compressed and the inner arc is excessively open, it is most likely to result in elongational and rheological behavior of bolts, while the opening of the gasket is very small. Although it will not lead to water stopping failure, chances are that the outer arc of the joint will crush. However, when the inner arc is compressed and the opening of the outer arc reaches 1 mm, the opening of gasket is 0.843 mm, the bolt will only elongate 0.357 mm; when the opening of outer arc is up to 4 mm, the bolt is elongated by 1.428 mm, and the opening of the gasket is 3.371 mm. According to the waterproofing requirements, when the outer arc is open by 6 mm, the opening of gasket is 5.057 mm, and the bolt is elongated by 2.142 mm. Therefore, when the inner arc is in the state of compression, the elongation of the bolts is relatively small. Generally, elongational and rheological behavior of bolts is not likely to occur, but is most probably to result in the water-stop gasket failure.

There are two types of longitudinal openings of the segment joints, including an initial deformation caused by the untightened bolts, and an elastic–plastic deformation of the bolts and segments when the tunnel is compressed. If all of the aforementioned joint opening is caused by plastic deformation of the bolt, the strain of the bolt is far more than 1%, and the bolt is already in the elongational and rheological state. The increment ΔD of the transverse diameter of the tunnel in this state has exceeded 10 cm compared to the perfect circle and has approached $1.6\%D$.

According to the existing circumferential structure and actual deformation of Shanghai Metro Line 1, the safety conditions for lateral deformation of the tunnel can be obtained: no tensile and compressive damage of the segment, i.e., no longitudinal cracks, good roundness of the lining ring assembled from the segments, no water seepage in the longitudinal joints of the assembled lining ring. Since there are tenon and mortise grooves between the ring joints, the lining ring can maintain its structural stability under the action of the pressure from the surrounding soil layer even if there are no circumferential bolts. Generally speaking, if the lateral deformation of the tunnel ΔD does not exceed $2\%D$, the tunnel structure remains safe.

At present, there is a lack of understanding of the development of deformation in operational tunnels at home and abroad. What's more, the control standards for permissible deformation are not available. There has been no research on the development of internal stress and structural safety of tunnels based on the parameter of diameter variation of the tunnel. The safety control standards, based on the deformation of the tunnel diameter, have the characteristics of being intuitive and easy to operate, which is of great significance for the protection of the structural safety of the operational tunnels.

2.4.3 Numerical Simulation of Tunnel Deformation Control

1. Numerical Analysis Model

In order to analyze the influence of tunnel load and lateral resistance on the lateral deformation of the tunnel, the straight-joint shield tunnel of Shanghai metro is taken as case study, and considering the influence of tunnel load and lateral compression coefficients of the soil, a load-structure calculation theory is proposed that the tunnel structure can be computationally analyzed by using the ABAQUS numerical simulation method. The segments and coupling bolts are entity units, the joints are of hard contact surfaces. The diameter of the bolt is 30 mm, and diameter the bolt hole is 42 mm, soil resistance is simulated by springs, and the hyperbolic constitutive model is adopted for concrete and bolts. The numerical calculation model, stratum resistance and load distribution are shown in Fig. 2.25, and the calculation parameters are listed in Table 2.6.

2. Calculation of Working Conditions

The numerical calculation model takes into account different loads, different lateral pressure coefficients and soil resistance coefficients, etc., and the range of compression load is 0 ~ 60. The lateral pressure coefficients are 0.5, 0.6, 0.7, and 0.8, and the soil resistance coefficients are K_s = 625, 1250, 2500, 5000, and 10,000 kPa/

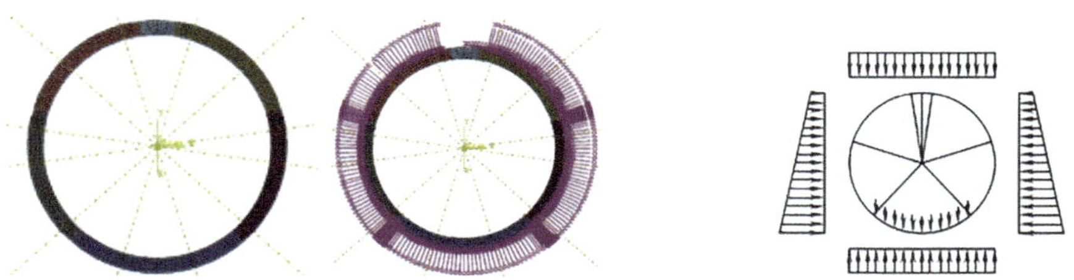

(a) Numerical calculation model (b) Stratum resistance (c) Load distribution

Fig. 2.25 Numerical calculation model and load distribution

Table 2.6 Numerical calculation parameters

Material	Modulus of elasticity of the first stage (MPa)	Modulus of elasticity of the second stage (MPa)
Concrete	35,495	592
Bolt	2.06×10^5	7×10^4

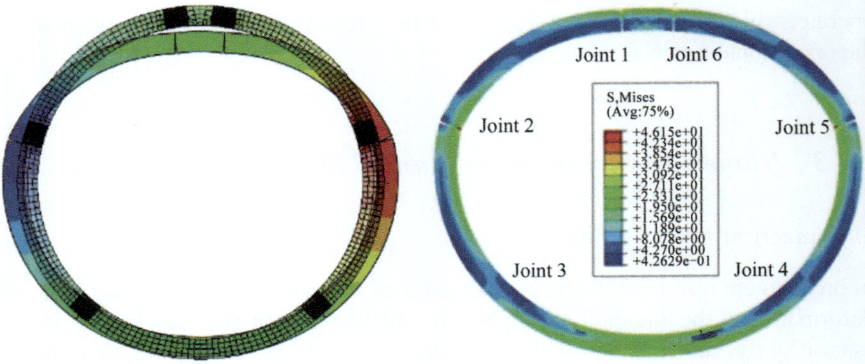

Fig. 2.26 Characteristics of transverse deformation and concrete force in tunnels under load

m, respectively. By combining the load, the coefficient of static lateral pressure on the soil and the coefficient of resistance of the soil stratum, a total of 29 operating conditions have been calculated, taking into account the situation when the joint bolts reach their ultimate stress and fail to function.

3. Lateral Deformation Analysis

The tunnel deformation and concrete forces under the upper load are shown in Fig. 2.26.

Significant lateral deformation will occur as a result of load. In the process transverse deformation, the structural deformations and forces develop symmetrically due to the symmetrical nature of the straight-joint tunnel. When the deformation occurs, the inner side of joints 1 and 6, open, and the outer side of joints 2 and 5 open at the same time, and the outer side of joints 2 and 5 open the widest, followed by inner side of joints 1 and 6, and the opening of joints 3 and 4 are the smallest. Due to their locations, the bolts of joint 1 and joint 6 are subjected to higher stresses, while larger concentrations of concrete stresses occurs near the opening joint 2 and joint 5.

Numerical analysis shows that the transverse deformation characteristics of the tunnel are significantly influenced by the magnitude of the load, the soil resistance coefficient and the soil lateral pressure coefficient. For a certain transverse pressure coefficient, the effect of load on tunnel deformation and force is shown in Fig. 2.27.

The effect of soil resistance coefficients on tunnel deformation and forces for a given lateral pressure coefficient is shown in Fig. 2.28.

The effect of soil lateral pressure coefficients on tunnel deformation and forces is shown in Fig. 2.29.

The bolt will break down when it reaches its ultimate limit, and the tunnel deformation after the bolt breaks down is shown in Fig. 2.30.

From the results of the numerical analysis above, it is found that with a constant lateral pressure coefficient, as the load on the top of the tunnel increases, the lateral deformation of the tunnel increases, the stresses on the bolts and the concrete increase, and the joints appear to accelerate in tension. The effect of the soil resistance and

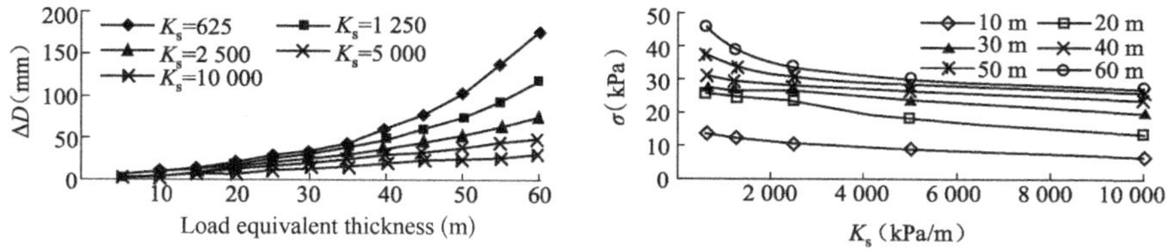

(a)Effect of load on tunnel diameter

(b)Effect of load on the maximum stress of the concrete

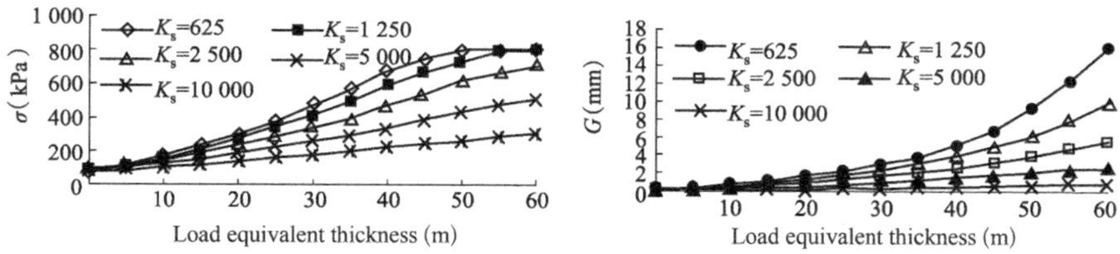

(c)Effect of load on the maximum stress of the bolt

(d)Effect of load on the maximum joint opening

Fig. 2.27 Effect of load on the deformation and forces at a lateral pressure coefficient of 0.7

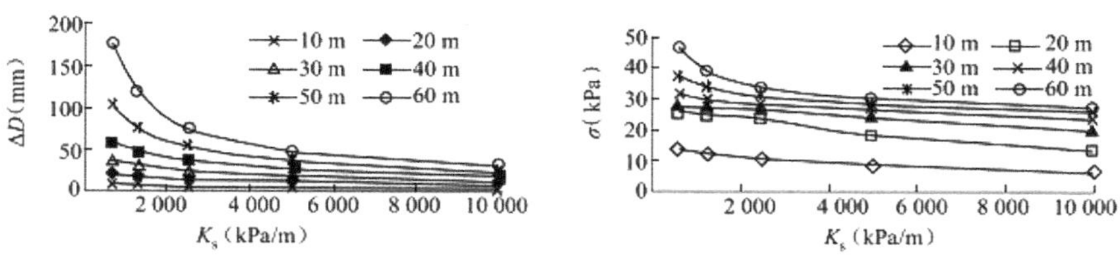

(a) Effect of the soil resistance coefficient on the variation of maximum diameter

(b) Effect of soil resistance coefficient on the maximum stress of concrete

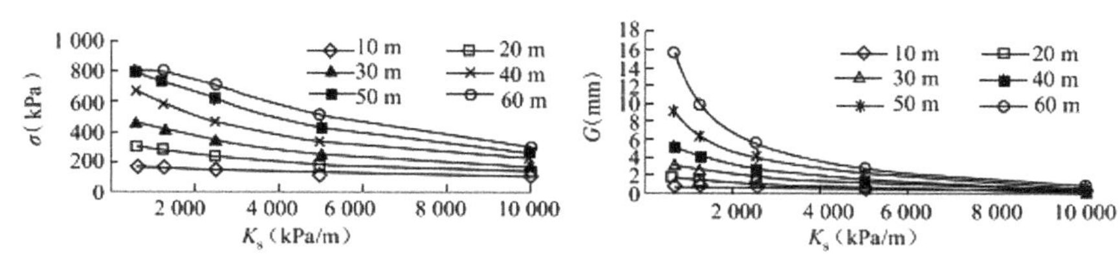

(c)Effect of the soil resistance coefficient on the maximum stress of a bolt

(b) (d)Effect of soil resistance coefficient on the maximum joint opening

Fig. 2.28 Effect of its resistance coefficient on the tunnel deformation and forces when the soil lateral pressure coefficient is 0.7

lateral pressure coefficients becomes more significant when the load on the top of the tunnel exceeds the load level corresponding to a 10 m underground depth.

According to the numerical simulation results, when the transverse deformation of the tunnel reaches 50 ~ 60 mm (0.8%D ~ 1%D), the bolts at joint 1 yield first, and the opening at joint 2 reaches the design control standard 5 mm. When the transverse

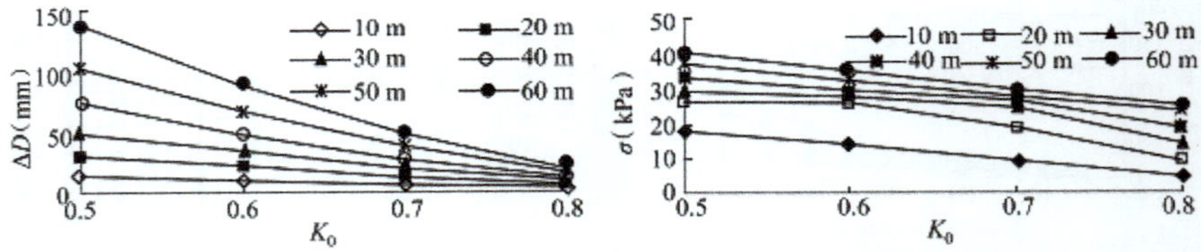

(a)Effect of soil lateral pressure coefficient on the variation of maximum diameter

(b)Effect of soil lateral pressure coefficient on the maximum stress of concrete

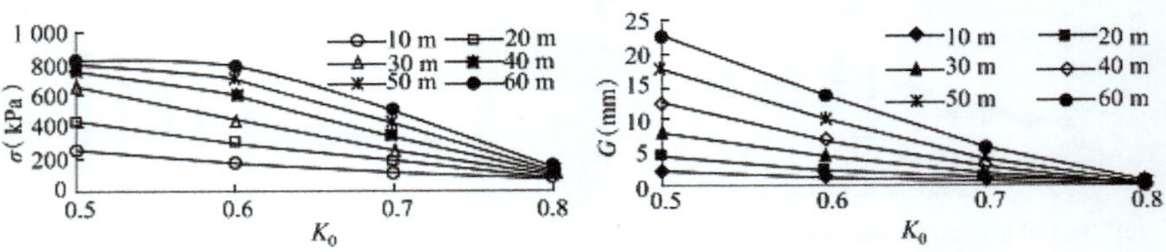

(a) Effect of soil lateral pressure coefficient on the maximum stress of a bolt

(b) Effect of soil lateral pressure coefficient on the maximum joint opening

Fig. 2.29 Effect of soil resistance coefficients at 5000 kPa/m on tunnel deformation and forces

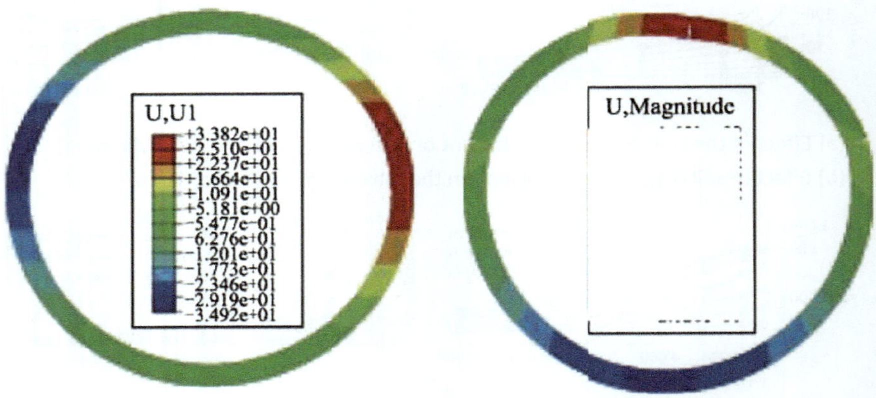

(a) Transverse deformation of the bolt before breakage (b) Transverse deformation of the bolt after fracture

Fig. 2.30 Effect of bolt fracture on transverse deformation of the tunnel

deformation develops to 80 ~ 95 mm (1.3%D ~ 1.5%D), the bolts at joint 1 reach ultimate strength, the concrete in joint 2 reaches the strength standard value, and the opening of joint 2 continues to increase to 7 ~ 8 mm; when the transverse deformation continues to develop to 120 ~ 150 mm (1.9%D ~ 2.4%D), the bolts in joint 2 yield, the concrete in joint 1 reaches the strength standard value, and the opening of joint 2 reaches 15 mm; when the convergence deformation of the tunnel increases to 212 mm (3.4%D), the bolts in joint 2 reach the ultimate strength.

If the bolt of joint 1 breaks, the convergence deformation of the tunnel increases dramatically by about 3 times at the moment of breakage, the concrete stress of joint 2 increases double, the opening of joint 1 increases by 5–6 times, the opening of joint 2 increases by about 3 times, and the bolt stress is transferred to the bolts of joint 2; the convergence deformation of the tunnel increases further, and at this moment, the bolt stress is transferred to the bolts of joint 5, and finally, the structure is damaged due to the excessive deformation of the structure.

The upper load will result in the tensile failure of the top bolts first, followed by compression damage to the springline concrete. At the moment the bolt break, the diameter of the tunnel increases dramatically by about three times.

From the above calculation and analysis, we know that when the soil cover is shallow, the transverse deformation of the tunnel increases linearly as the upper load increases, and the smaller the soil resistance coefficient is, the larger the transverse deformation of the tunnel is; when the stress of the concrete of the segment reaches the maximum, the tension of the bolt and the opening angle of the joint increase linearly in accordance with the underground depth of the tunnel; the transverse deformation decreases rapidly as the soil resistance coefficient increases, and the variation of the maximum inner force of the concrete and the bolts is not significant as the soil resistance coefficient increases; the larger the soil resistance coefficient is, the smaller the joint opening is; the transverse deformation, the inner force of the concrete, bolt tension and joint opening decrease significantly as the lateral pressure coefficient of the soil increases.

The above analysis shows that the soil resistance coefficient, the lateral pressure coefficient and the magnitude of the load are interacted, determining the tunnel's stability. The load on the tunnel is detrimental to the stability of the tunnel, and an increase in the lateral pressure coefficient and the resistance coefficient of the soil stratum are beneficial to the stability of the tunnel.

2.4.4 Simplified Geometric Analysis Method for Tunnel Transverse Deformation

1. Principles of Simplified Geometric Analysis Method

Numerical simulations have shown that variations in the magnitude of the load, soil lateral compression coefficients and soil resistance coefficients do not affect the relative spatial relationships between the tunnels, although they do cause changes in the tunnel diameters, concrete and bolt stresses, and joint openings. According to this feature, it is required that a stable structural force system between the segments should be kept after a certain amount of deformation has occurred to maintain the stability of the structure in the deformation process. In view of the symmetry of the tunnels, a simplified geometric analysis method for the straight-joint tunnels is proposed (Wang and Zhang 2013).

Considering the symmetry of the transverse deformation of the tunnel, the simplified geometric analysis method focuses on analyzing the relative rotation and horizontal displacement between the segments, assuming that the segment at the bottom of tunnel does not move, and the transverse deformation of the tunnel occurs under the influence of the load. The tunnel will remain a hexagonal shape with left–right symmetry after deformation due to the rigid body displacement and rotation of the segments, as shown in Fig. 2.31. The deformation process is shown in Fig. 2.32.

Based on the characteristics of the transverse deformation development as illustrated in Figs. 2.31 and 2.32, taking the vertical displacement *d* (variation in the vertical diameter of the tunnel) of the capping segment as a variable, the development of the tunnel diameter and the joint openings is analyzed and explored. According to the geometric relationship between the tunnel segments, the correlation of the deformation at each characteristic point of the tunnel and the displacement of the capping segment is analyzed as shown in Table 2.7. Each deformation indicator in Table 2.7 is illustrated in Fig. 2.33.

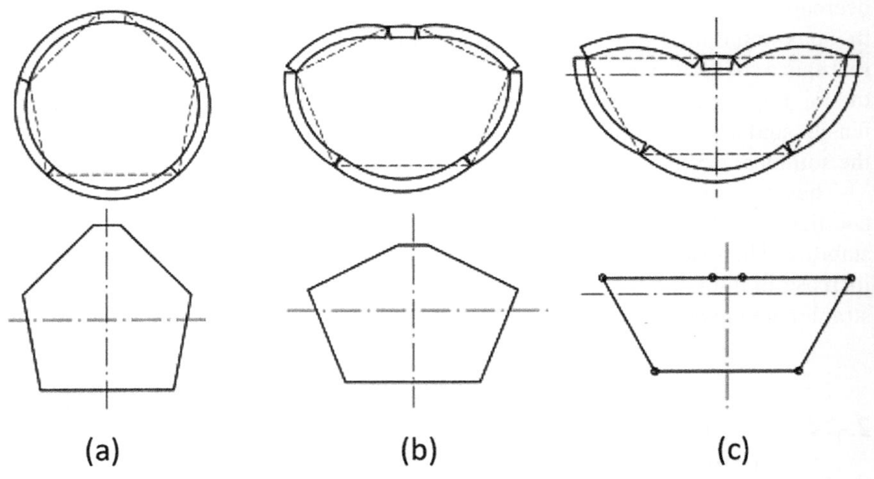

(a) simple graph to the tunnel before deformation
(b) simple graph to the tunnel after deformation
(c) ultimate flattening state

Fig. 2.31 Simplified geometrical analysis method for transverse deformation in tunnels

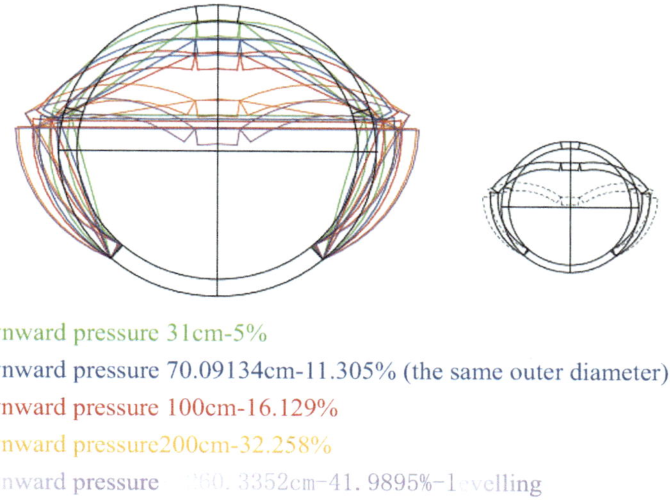

Fig. 2.32 Diagram of progressive development of tunnel deformation

Table 2.7 Relationship between tunnel deformation and crown settlement

Parameter for tunnel deformation	Calculation equation
θ_1	$\arccos\left(\frac{A^2}{6.3141A}\right) - \arccos\left(\frac{1.6429}{A}\right) + 45.8651°$
θ_2	$125.73° - \arccos\left(\frac{10.7474d - d^2 - 11.6407}{19.9337}\right)$
θ_3	$\arccos\left(\frac{A^2}{6.3141A}\right) + \arccos\left(\frac{1.6429}{A}\right) - 100.1349°$
l_1	$0.7 \cdot \cos\left[52.1349° - \arccos\left(\frac{A^2}{6.3141A}\right) + \arccos\left(\frac{1.6429}{A}\right)\right] + 0.8629$
l_2	$6.3141 \cdot \sin\left[90° + \arccos\left(\frac{A^2}{6.3141A}\right) - \arccos\left(\frac{1.6429}{A}\right)\right] + 0.8628$
l_3	$4.8016 \cdot \sin\left[\arccos\left(\frac{A^2}{6.3141A}\right) + \arccos\left(\frac{1.6429}{A}\right) - 83.7884°\right] + 4.1486$
l_4	$4.1486 - 0.7 \cdot \cos\left[\arccos\left(\frac{A^2}{6.3141A}\right) + \arccos\left(\frac{1.6429}{A}\right) - 52.1349°\right]$
	$A = \sqrt{31.5744 - 10.747d + d^2}$

2. Law of Deformation Development

(1) Law of Joints Opening Development

According to the relationship between the tunnel deformation index and the tunnel vertical diameter variable as shown in Table 2.7, the opening of the tunnel joints can be obtained as the tunnel vertical diameter varies, as shown in Fig. 2.34.

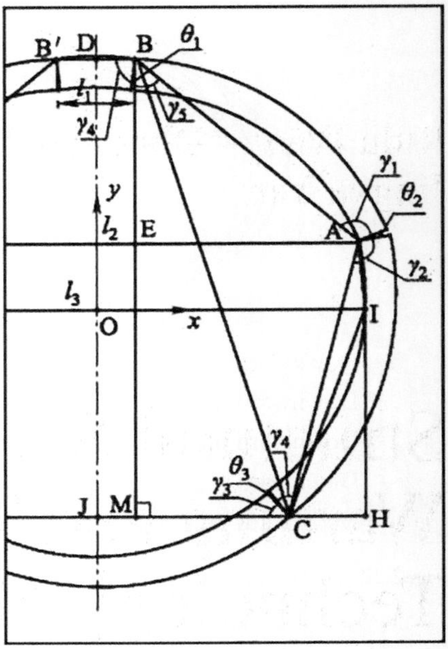

Fig. 2.33 Diagram of the physical significance of tunnel deformation indicators

From Fig. 2.34, it can be seen that the tunnel joint opening in relation to the vertical diameter variation exhibits the following pattern.

① In the process of tunnel deformation, the opening of the outer side of the tunnel joint at the springline is equal to the sum of the opening of the inner side of the tunnel joints at the crown and intrados, i.e., $\theta_2 = \theta_1 + \theta_3$ and this relationship is always satisfied in the process of tunnel deformation. It can be seen that during the deformation of the tunnel, the joint opening at the tunnel springline is the most significant, and this pattern is consistent with the results obtained from the finite element calculation simulation.

② In the event of smaller deformations (vertical diameter variation of 5%D or less), θ_1, θ_2, θ_3 all increase linearly as the vertical diameter decreases.

③ Waterproofing failure is likely to occur since the opening of the joint at the springline θ_2 is relatively large. The opening often emerges on the outside of the joint and the corresponding inner wall is under significant pressure and susceptible to compression damage.

④ From the relationship between the joint deformation, it can also be seen that θ_1 accounts for about 60% of θ_2, so F_2 the opening of the outer wall is more significant, as a result, the inner wall of the joint is under heavy pressure, this has also been verified by the numerical analysis.

⑤ When the tunnel is pressed down by almost 3 cm, θ_2 opens up by about 1°, and in accordance with the thickness of the segment, the joint between segments opens up by 6 mm, i.e., the external waterproofing is approaching the margin of

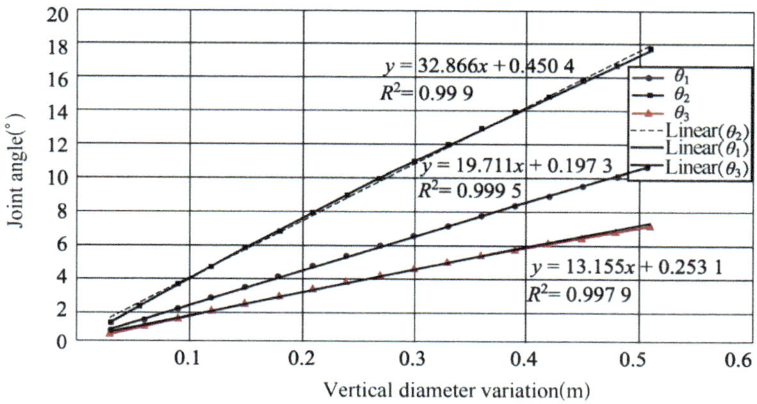

(a) The variation of the inner joint opening when the variation of vertical diameter is less than 50cm

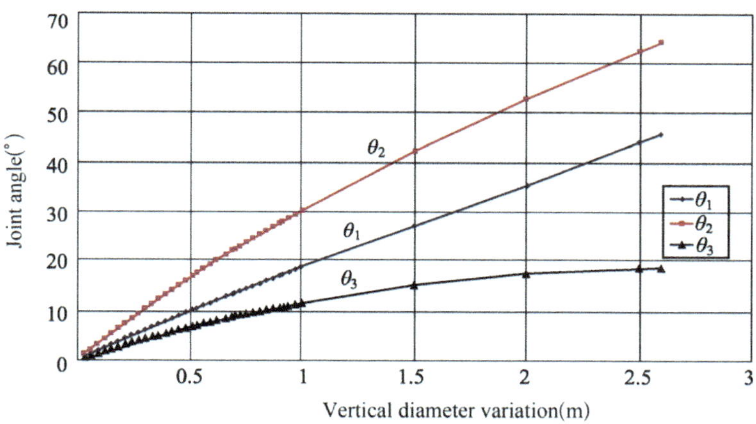

(b) The variation of the joint opening when the vertical diameter varies

Fig. 2.34 Patterns of change in tunnel joint tensioning with vertical diameter change

the design value, and the waterproofing capacity of the tunnel is still relatively vulnerable.

(2) The Development of Horizontal Diameter Deformation

In the process of tunnel deformation, the variation of the horizontal diameter in relation to the vertical diameter of the tunnel is shown in Fig. 2.35. From Fig. 2.35, it can be seen that the transverse deformation of the tunnel is a process in which the horizontal diameter decreases as the vertical diameter increases. l_2 is the most sensitive to variation of d, followed by l_3, l_1, and l_4 in sequence. The relationship between them can be described in a quadratic parabolic approximation. From Fig. 2.35, it can be found that the distance between neighboring segments l_2 reaches the diameter of the

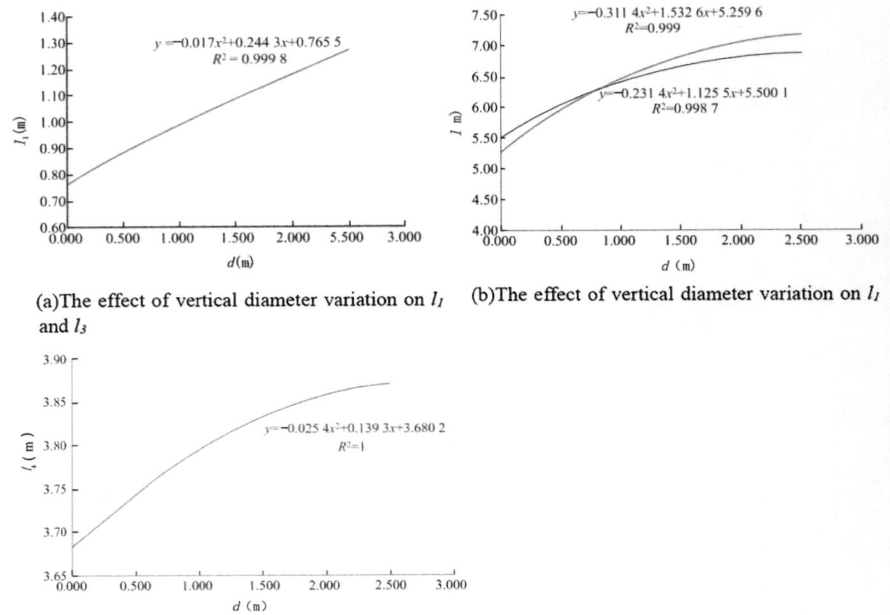

(a)The effect of vertical diameter variation on l_1 and l_3

(b)The effect of vertical diameter variation on l_1

(c)The effect of vertical diameter variation on l_4

Fig. 2.35 The variation pattern of horizontal diameter in relation to vertical diameter

tunnel when the vertical diameter of the tunnel decreases by about 0.7 m. When the reduction of the vertical diameter of the tunnel reaches 2.6 m, the top of the hexagonal shape will be flattened. Theoretically, the process of transverse deformation can be reversed (the deformation is irreversible if the variation of the vertical diameter has reached 2.6 m) but it requires variation of the external load. Table 2.8 shows the quantitative relation between the tunnel deformation index and the variation in vertical diameter. When vertical diameter decreases by less than 0.25 m, the tunnel deformation are highly consistent with vertical diameter variation compared to the finite element calculation.

3. Comparison Between Four Simple Analysis Methods

Four analysis methods of the transverse convergent deformation of the tunnel are compared, it can be found that

(1) Assuming that the tunnel segment is rigid, the relationship between the opening of each longitudinal joint of the tunnel and the variation in the distance between the points of interest in the tunnel is analyzed by the geometric analysis. The assumption is consistent with the results of the on-site full-scale ring test and the measured data, as shown in Fig. 2.36.
(2) According to the results of finite element calculation, the inner side of the joint between the capping segment and its neighboring segment θ_1 are open; the outer

Table 2.8 Relationship between the vertical and transverse deformation of the tunnel

Transverse radius increment ΔD	Vertical diameter reduction (cm)										
	3	6	9	12	15	18	21	24	27	30	35	70
	3.818	7.533	11.151	14.677	16.811	17.146	24.473	27.793	5.8105	34.106	39.037	69.011
θ_1	0.6560	1.3014	1.9376	2.5652	3.1849	3.7969	4.40193	5.0001	5.5920	6.17801	7.1421	13.5515
θ_2	1.1301	2.2392	3.3282	4.3981	5.4500	6.4848	7.50338	8.5065	9.4948	10.4692	12.0635	22.3656
θ_3	0.4741	0.9378	1.3905	1.8328	2.2651	2.6878	3.10145	3.5063	3.9028	4.2912	4.9214	8.8141

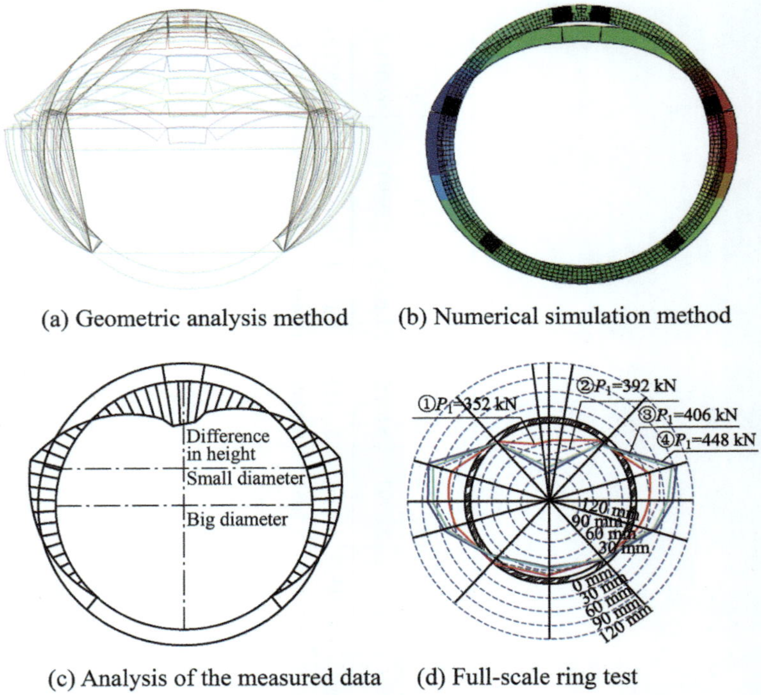

Fig. 2.36 Comparison between the results of 4 different analysis methods

side of the joint between the neighboring segment and the standard segment θ_2 are open; the inner side of the joint between the standard segment and the invert θ_3 are open, and θ_2 is the largest, followed by θ_1 and θ_3 in sequence. Numerical simulation and analytical methods have revealed that the variation of the horizontal diameter and the opening of the joint between the neighboring segment and the standard segment are linearly related to each other when the deformation is small, which has proven the rationality and correctness of the geometrical analysis.

2.4.5 Transverse Deformation Control

By numerical simulation of the development of large transverse deformation of the tunnel under load conditions and the simple geometrical analysis of the tunnel deformation, the development pattern of large transverse deformation of the straight-joint tunnel can be obtained, and it can be concluded that:

(1) Under the tunnel load conditions, the pressure on the extrados and invert of the tunnel is intensified, the extrados at the joints is closed under pressure and the

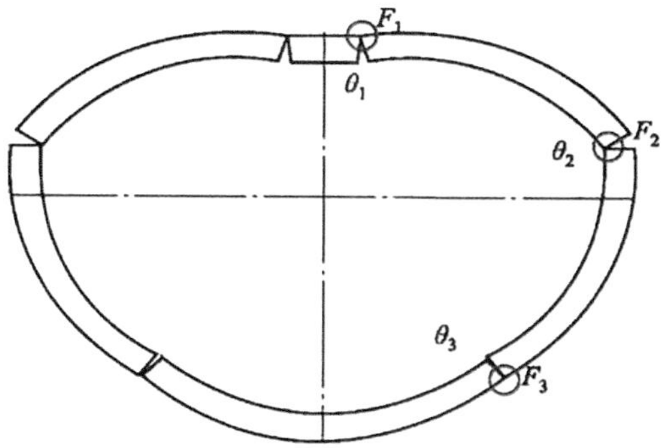

Fig. 2.37 Diagram of transverse deformation of the tunnel

intrados opens under tension; the inner wall joint at the top 0° point of the tunnel is open and the outer wall joint is closed under pressure, making it susceptible to compression damage; the inner wall joint at the 73° above the springline is closed and compression damage is likely to occur. The outer wall opening reaches its maximum, water leakage is most likely to occur; the joint between inner wall invert segment and the standard segment opens, the outer wall joint is subject to compression damage.

(2) The compression damage occurs in the following order: the inner wall at the springline > outer wall at the top of the tunnel; the opening follows the following order: outer wall at the springline > inner wall at the top of the tunnel; theoretically, water leakage is not likely to occur when the outer wall is compressed, but when the stress exceeds the compression strength of the concrete at the joints, the concrete at the joints is subject to compression damage.

(3) According to the finite element simulation calculation, the most serious compression damage occurs at the inner wall joint of the springline, followed by the outer wall of the top longitudinal joint, and then at the outer wall of the longitudinal joint of the invert.

(4) The largest opening is located at the outer wall of the springline, followed by the inner wall of the longitudinal joint at the top, and then at the inner wall of the invert, i.e., $F_{2\,\text{outer}} > F_{1\,\text{inner}} > F_{3\,\text{inner}}$, and the physical index of F is shown in Fig. 2.37. In the process of transverse deformation of the tunnel, different parts of the tunnel manifests different patterns of damages, the similar water leakage phenomenon may occur in different parts of the tunnel, but the degree of danger varies. Special attention should be given to the water leakage occurring at the compressed parts of the tunnel.

(5) For a constant vertical/horizontal stress ratio, the greater the transverse deformation, the greater the lateral resistance of the tunnel, and the more likely the compression damage will occur; for an increasing vertical/horizontal stress ratio, the tunnel is more likely to collapse.

References

Ministry of Housing and Urban-Rural Development of China (2008) Code for construction and acceptance of shield tunneling: GB 50446-2008. China Architecture & Building Press, Beijing

Wang R, Zhang D (2013) Study on lateral deformation mechanism and control indexes of soft soil shield tunnel under overload. Chin J Geotech Eng 35(6):1092–1101

Wang R, Jia J, Liao S (2013) Shanghai metro monitoring practice. Tongji University Press, Shanghai

Chapter 3
IntelliSense Technology of Tunnel

The soft soil layer in Shanghai has a high void ratio, compressibility, water content, and sensitivity, but low shear strength. Disturbance significantly reduces its strength, leading to prolonged consolidation and secondary consolidation settlement over several years. The tunnel structure laid in this kind of layer is affected by construction quality, storage conditions, dynamic and static loads during operation, and maintenance quality. As the service life increases, long-term issues such as overall settlement, differential settlement, and convergence deformation of the tunnel cross-section can lead to tunnel diseases worsening, such as structural damage, deformation, and water leakage. Meanwhile, construction activities around the tunnel intensify the development of structural defects, with these issues interacting and exacerbating each other. If left unmanaged and allowed to progress, they will inevitably threaten operational safety.

The prerequisite for controlling and treating tunnel diseases is the evaluation of the structure's condition and the severity of the damages, which is based on the perception of the tunnel's structural status. Therefore, adopting IntelliSense technology for effective monitoring of subway tunnels is key to ensuring structural and operational safety. In this chapter, we introduce the main perception methods of structural damages based on more than 20 years of structural maintenance practice in Shanghai: (1) Manual long-term monitoring (aiming at long-term settlement, long-term convergence), regular inspections on structure; (2) Immobile automatic monitoring (EL beam sensor, hydrostatic level, laser rangefinder, inclinometer, surveying robot, Fiber Bragg Grating, etc.); (3) Mobile monitoring (laser scanning-based mobile detection and camera group-based mobile damage detection). IntelliSense mainly includes appearance inspection and measurement of the inner wall of the tunnel, detection of the interior of the pipe segment, and gap detection behind the tunnel wall. The pipe segments are precast in factories, ensuring the highest quality, particularly in strength and dimensional accuracy. Special equipment has been developed for the

R. Wang et al., *Structural Safety Early Warning and Treatment Technology of Metro Tunnel in Soft Soils*, Key Technologies for Tunnel Construction under Complex Geological and Environmental Conditions,
https://doi.org/10.1007/978-981-96-6410-8_3

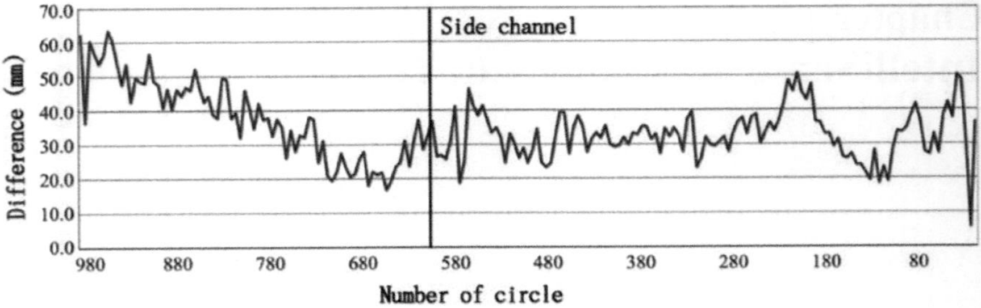

Fig. 3.1 The difference between the horizontal diameter and the design diameter of the station-to-station tunnel (Longyang-Zhangjiang) of Line 2 after eight-year running

detection of synchronous back filling grouting quantity and grouting quality. This chapter focuses on rapid perception and warning of damages and deformations on inner wall appearance of tunnel.

3.1 Types and Monitoring System of Tunnel Structural Diseases

3.1.1 Main Types of Structural Deformation

Structural deformation mainly includes longitudinal settlement and transverse convergence. The value of settlement in some sections of Line 1 tunnel reached 29.15 cm, as shown in Fig. 3.1; Fig. 3.1 is the difference between the horizontal diameter and the design diameter of the station-to-station tunnel (Longyang-Zhangjiang) of Line 2 after eight-year running, with convergence deformation of many cross-sections being larger than 5‰*D* (*D* is the design external diameter of the tunnel).

3.1.2 Main Types of Surface Diseases on Structure

Surface diseases on structure include water leakage, structural damage, dislocation, trackbed disengagement, etc. (Guo 2014).

Water leakage, usually occurring in joints of segments, grouting holes, bolt holes, cracked pipe segments, etc., is classified into four types according to the size of the seepage halo, as shown in Table 3.1.

Structural damages refer to crack, corner spalling, as shown in Table 3.2.

Other structural damages include the disengagement between trackbed and segments, segments dislocation, expansion of joint, as shown in Table 3.3.

Table 3.1 Degree and expression of water leakage

Types	Symbols	Description of symbols	Notices
Water tracks		A closed curve filled with dashed lines	The curve boundary should be determined by the actual water tracks distribution
Water seepage		A closed curve filled with slashes	The curve boundary should be determined by the actual water seepage distribution
Dripping	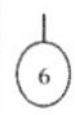	It consists of three parts: vertical bar, ellipse, and number indicates the drip frequency (dripping number/min)	1. When less than 1 drop/min, < 1 should be marked inside the ellipse 2. When greater than 60 drops/min, it can be considered that drip has formed a line flow, and it should be marked according to the percolation flow
Sand and mud leakage		A closed curve filled with points and small triangles	The curve boundary should be determined by the actual boundary

Table 3.2 Degree and expression of structural damages

Types	Symbols	Description of symbols	Notices
Crack		Curve or polyline is based on the actual line shape of the crack	The crack width should be remarked which can be measured
Missing pieces		The boundary is based on the actual missing pieces boundary	The range of actual missing pieces should be completed, and the depth of the missing pieces should be remarked, which can be measured
Defect		Straight segment represents the longitudinal where the defect occurred	–

Table 3.3 Degree and expression of structural deformation

Types	Symbols	Description of symbols	Notices
Segments dislocation		The straight line is perpendicular and crosses to the joint at the staggered table, and the number indicates the amount of staggered table	The numbers should be recorded on the opposite side toward the recorder
Separation between roadbed and segments		The symbol is recorded at the junction between the roadbed and the segment	–
Expansion of joint		Inverted V-shaped, representing the joint has expanded	–

3.1.3 Monitoring and Maintaining System for Tunnel Structure

The underground nature of tunnels complicates the identification of unforeseen issues, hindering our full understanding of the safety status of subway tunnel structures. Therefore, ensuring absolute safety through the development and enhancement of monitoring, maintenance, and evaluation systems, along with a reliable warning system, is crucial. Over the past 20 years, the Shanghai Metro has implemented the following measures, resulting in a series of workplace rules and regulations.

(1) Regular inspections on tunnel and protected areas along the line *Customized inspection methods and Standardized Operation Rules for Disease Inspection in Shanghai Rail Transit Tunnels* have been established and implemented. According to the Rules, measures including a half-month inspection on tunnel structure, ground patrol along the line for detecting and stopping illegal operations promptly have been implemented. Also, to strengthen technical review and on-site monitoring of various operation plans near the tunnel for strictly controlling construction disturbances along the subway tunnel.

(2) A permanent subway monitoring system was established, long-term monitoring on structure and engineering impacts was regularly implemented to ensure the continuity and accuracy of monitoring data. To closely monitor the settlement displacement and convergence deformation by establishing permanent layered settlement monitoring points along the subway line. As for inter-zone tunnels with unstable settlement and deformation, attention should be paid from the strategic perspective of subway safety, and the trend of deformation should be predicted to ensure that tunnels are under control.

① Long-term and regular monitoring on running lines (a) the line that was put into use newly requires settlement monitoring four times a year and convergence monitoring once a year; (b) the line that was more than two-year running requires settlement monitoring two or three times a year and convergence monitoring once a year; (c) intensified monitoring should be implemented for sections showing abnormalities. Subway running in a safe and normal way is a long-term task, which means that inspection work needs to be carried out continuously and uninterrupted, the road network has established a unified subway horizontal control network and a settlement control network based on bedrock markers.

② Monitoring on engineering projects within the protected zone. The monitoring frequency of structural deformation resulting from engineering projects depends on the project type's risk level and construction phase, with high-risk projects requiring real-time, online monitoring. Normal indicators of monitoring include convergence, displacement, settlement, other special indicators, like stress, strain, and crack, can be added according to projects.

(3) Long-term monitoring of the tunnel, along with inspection and monitoring activities, yields an annual analysis report for the Shanghai Metro, highlighting tunnel diseases and issues requiring resolution or attention. Additionally, a medical record card system should be established for long-term tracking and control toward major diseases.
(4) Based on the tunnel inspection and analysis of existing data, it is necessary to conduct targeted tracking and monitoring analysis of key sections of the tunnel. Especially 5 for sections, curves, and sections with large settlement that have been impacted by engineering activities, key monitoring and measures should be taken promptly if any abnormalities are found.
(5) To establish and improve metro files about monitoring and maintaining, which includes design and completion data, construction technology, equipment, grouting materials, etc. during the construction period; information on inspection and repair of tunnel diseases, long-term deformation data of the tunnel (long-term settlement of roadbed, tunnel displacement, pipe diameter convergence), and engineering activity data within protected zone during the operation period.

3.2 Routine Monitoring and Inspection

Routine monitoring and inspection primarily involve manual operations are conducted regularly, focusing on long-term settlement, long-term convergence, and structural disease inspections.

3.2.1 Long-Term Settlement Monitoring

Long-term settlement monitoring typically uses precision leveling, performed regularly to capture both staged and cumulative changes in the vertical displacement of rail transit structures. Attention would be paid to changes at the junction of different structures of the tunnel (station, end shaft, tunnel, air shaft, etc.) and the weak points of the shield tunnel (side channel, accident section during construction, etc.), and to find out the abnormal conditions of the tunnel in time. Based on the different impacts of these changes on rail transit, different suggestions and measures are proposed for accumulating engineering experience and data for the subsequent design and construction of rail transit tunnels. Precision leveling mainly includes measuring vertical datum network and monitoring deformation points.

In addition to adhering to the Specifications for First and Second Order Leveling (GB/T 12897) and the Code for Deformation Measurement of Building and Structure (JGJ8), technological innovations in long-term settlement monitoring have been achieved by aligning with Shanghai's geological conditions and structural features.

(1) 45 groups of bedrock markers have covered Shanghai, unifying the benchmarking framework for settlement monitoring of Shanghai Metro network.
(2) Research on the deployment methods of monitoring networks, including bedrock markers, deep-buried benchmarks along the line and station benchmarks.
(3) Research on the observation methods of settlement point based on *i*-angle correction of level.

3.2.1.1 To Unify the Benchmarking Framework for Settlement Monitoring of Shanghai Metro Network

Shanghai rail transit has been under running and maintenance over years, because the settlement is high in frequency, normal benchmarks could not meet the requirements of monitoring and measuring. It is very important to set bedrock markers as the frame network for settlement measurement of Shanghai Metro, which covers 45 groups of bedrock markers in Shanghai, and the distribution of the whole bedrock markers network is shown in Fig. 3.2. The base bedrock marker is consolidated with the bedrock to ensure stability, reliability, and permanence.

3.2.1.2 Deployment Methods of Monitoring Networks, Including Bedrock Marker, Deep-Buried Benchmark Along the Line and Station Benchmark

The bedrock markers in Shanghai are generally distributed incompactly, so the inspection line is long, and the measurement error is easy to accumulate. For handling this issue, the vertical control network is arranged by levels, mainly including the ground benchmarking network measurement, the roadbed leveling measurement, and the vertical connection measurement, as shown in Fig. 3.3.

(1) The elevation starting point of the benchmark network is based on the bedrock markers of Sheshan in Shanghai and the bedrock points distributed along the line. The leveling line of the first-class benchmark network should be laid out along highways and urban roads, and the leveling line should be closed into a loop or form an attached leveling line.
(2) To set up a deep-buried benchmark near each station as the working base point, and the base of the deep-buried benchmark should be deep into the second aquifer over 3 m.
(3) The settlement monitoring of each section and station starts and ends at deep level points, and a second-class reference network is set up.

The measurement accuracy of the elevation control network must meet the requirements of Table 3.4.

The technical requirements for elevation control surveying must meet the requirements of Table 3.5.

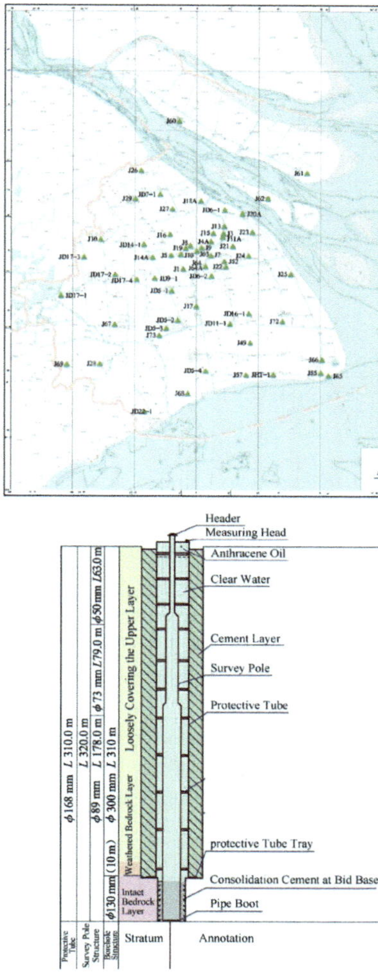

Fig. 3.2 Distribution diagram and structure of subsidence monitoring base point (bedrock) in the Shanghai Metro network

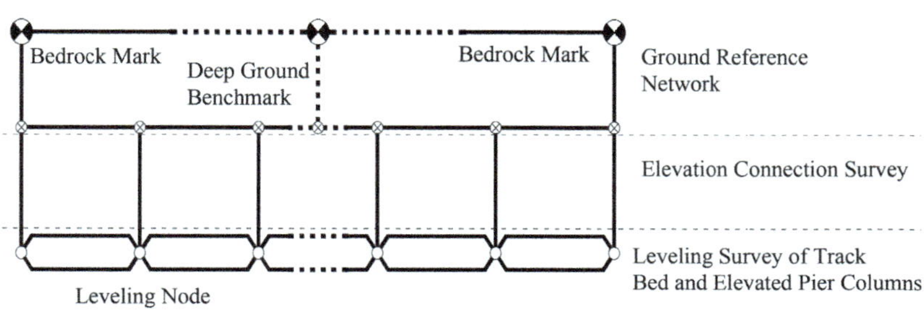

Fig. 3.3 Diagram of elevation control network layout

Table 3.4 Elevation control measurement accuracy requirements

Type	Accuracy category	Accuracy index (mm)
Ground reference network measurements	Incidental medium error in leveling per kilometer	± 0.45 ± 0.45
	Total median error of leveling per kilometer	± 1.0
Roadbed leveling	Incidental medium error in leveling per kilometer	± 1.0
	Total median error of leveling per kilometer	± 2.0
Elevation linkage measurements	Medium error in elevation difference between leveling points	± 1.0
	Error in station height difference	± 0.3

Table 3.5 Technical requirements for elevation control measurement

Type	Accuracy category	Accuracy index (mm)
Ground reference network measurements	Poor round trip, attachment or loop closure	$\pm 2.0\sqrt{L}$
	Detecting the difference between measured elevation differences	$\pm 3.0\sqrt{L}$
Roadbed leveling	Poor round trip, attachment or loop closure	$\pm 4.0\sqrt{L}$
	Detecting the difference between measured elevation differences	$\pm 6.0\sqrt{L}$
Elevation linkage measurements	Poor round trip, attachment or loop closure	$\pm 0.3\sqrt{n}$
	Detecting the difference between measured elevation differences	$\pm 0.4\sqrt{n}$

3.2.1.3 The Observation Methods of Settlement Point Based on *i*-Angle Correction of Level

The *i*-angle of the level is the main source of error, which significantly impacts settlement measurement. In theory, it can be minimized by having equal sight distance between the foresight and backsight. However, due to the ultra-long linear structure of rail transit, and the high density of measurement points, settlement measurement is carried out within a limited time window of each day, making it difficult to ensure equal visual distance. The influence of sight distance difference and *i*-angle of the level on elevation height can be calculated by using Formula 3.2.1.

$$\Delta h = \frac{\Delta D \times i}{\rho} \tag{3.2.1}$$

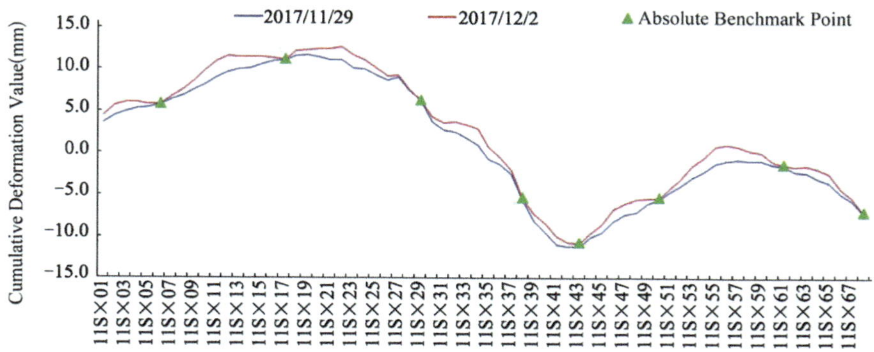

Fig. 3.4 Example of *i* angle influence on subsidence measurement curve

where

Δh Elevation correction factor.
ΔD Difference in sight distances between the middle and rear sight points where $\Delta D = D_{\text{middle sight}} - D_{\text{rear sight}}$.
i Leveling angle.
ρ Conversion constant between seconds and length, $\rho = 206{,}265$.

According to Formula 3.2.1, the maximum difference of sight distance is 31 m in practice. If 5″ of *i*-angle affects the result by 0.8 mm, 10″ would affect by 1.5 mm, and 15″ would affect by 2.2 mm. Figure 3.4 shows an example of the influence of *i*-angle of the level in settlement measurement, manifested as a wave shape related to the measuring station. At the same time, factors such as temperature, light, or large vibrations, would easily change the *i*-angle. Fixed instruments and fixed station positions cannot completely eliminate the influence of *i*-angle. Therefore, the *i*-angle of an electronic level should be checked each time before use to correct the impact of *i*-angle on the observation results. During each observation, information such as average temperature, external environment, weather conditions, should be recorded to facilitate comprehensive analysis of settlement measurement data.

3.2.2 Long-Term Convergence Monitoring

In Shanghai, shield tunnels mostly used continuous seam assembling, with a small number of tunnels using staggered joint assembly, and some sections using double circular staggered joint assembly. According to the general requirements of shield tunnel design: when the tunnel is assembled into a ring, under external load, the longitudinal and circumferential joints are allowed to open 6 mm without water leakage, and the cumulative change in diameter is less than 5‰*D* (*D* is the outside diameter of the shield tunnel). Therefore, for a single circular tunnel with an external diameter of less than 6.2 m, its deformation should be controlled within 3.1 cm under load. However, due to the complex and diverse factors during tunnel operation, the deformation of the tunnel often exceeds this control standard. Excessive deformation would cause the tunnel joints to open or compress, resulting in water leakage, detachment of the roadbed and segments, and even serious structural diseases such as segment damage due to pressing and bolt failure. Through long-term monitoring and treatment of soft soil tunnels, it has been found that tunnel diseases and structural deformation are closely related, which means that once the transverse deformation of the tunnel exceeds a certain limit, it not only causes leakage in joints but also serious pressure damage to the segments, and in extreme cases, it can even lead to the collapse of the entire tunnel.

Long-term convergence measurement is an important measurement item for rail transit tunnel monitoring, which uses instruments such as total station, distance measurement equipment, laser scanner, and inclinometer to obtain tunnel diameter and its changes. Based on the characteristics in structure and deformation of the Shanghai Metro tunnel, the following technological innovations have been made:

(1) The feasibility of using reflectorless distance measurement in convergence monitoring was studied, and an additive constant correction method was established.
(2) We have studied a full section scanning convergence monitoring method based on an automatic total station.

3.2.2.1 Convergence Monitoring Based on Reflectorless Distance Measurement

1. Feasibility study on using reflectorless distance measurement for long-term convergence measurement.

When using traditional convergence gauge, it is necessary to set up a convergence hook on the tunnel structure, and then install a convergence meter, and apply a fixed tensile force (usually 20 kg) during observation. For shield tunnels, traditional convergence measurement operations are difficult. In recent years, reflectorless distance measurement is available, therefore, accuracy analysis and verification have been carried out.

The experiment used 10 total stations on a measuring line of approximately 7.2 m to conduct a comparison test on reflectorless distance measurement. Each instrument shall be erected twice, and each station shall observe 10 times with a positive and negative mirror. The statistical data is shown in Table 3.6.

From the data in the table, it can be seen that:

(1) For the same baseline, the average measurement value of the 10 total stations used in the experiment ranges from 7.2047 to 7.2129 m, with a difference of 8.2 mm. This indicates that without correction for the short range constant difference, the accuracy cannot meet the accuracy requirement of long-term convergence measurement of ± 3 mm.

Table 3.6 Data statistics of 40 measurements each based on uncooperative target boiling distance using 10 total stations

Serial number	Instrument model	Comparison of 40 observations from the same instrument					Difference between the mean of each station and the mean of all observations (mm)
		Average (m)	Maximum values (m)	Minimum values (m)	Maximum and maximum smaller values are worse (mm)	Standard deviation (mm)	
1	TS30	7.2096	7.2106	7.2084	2.1	0.49	0.56
2	TC402	7.2057	7.2071	7.2045	2.6	0.57	− 3.34
3	TS15	7.2090	7.2103	7.2079	2.4	0.58	− 0.04
4	TS06	7.2120	7.2129	7.2115	1.4	0.34	2.96
5	TS06	7.2092	7.2107	7.2077	3.0	0.72	0.16
6	TC802	7.2129	7.2142	7.2105	3.7	0.76	3.86
7	TC1201	7.2047	7.2055	7.2040	1.5	0.29	− 4.34
8	TC1201	7.2093	7.2103	7.2078	2.4	0.54	0.26
9	TC1201	7.2086	7.2092	7.2072	2.0	0.40e	− 0.44
10	TC1201	7.2109	7.2118	7.2098	1.9	0.51e	1.86

(2) 40 observations arousing from a single total station are compared with a maximum standard deviation of ± 0.76 mm. After correcting the short range constant difference, the method based on reflectorless distance measurement can meet the accuracy requirement of long-term convergence measurement of ± 3 mm.

(3) During the annual calibration of instruments, it is generally not necessary to conduct calibration of additive constant in reflectorless distance measurement. Therefore, it is required that the operating unit should conduct a self-inspection before use to determine the detection of the correction constant for the short range of reflectorless distance measurement.

(4) The internal comparison of 40 observation values of each instrument shows that the difference between the maximum and minimum values also varies from 1.4 to 3.7 mm, indicating that the accidental error of a single measurement has a significant impact on the accuracy of long-term convergence measurement ± 3 mm. It is required that long-term convergence measurement should be observed in one measurement cycle.

Based on the above analysis, the research results suggest that a long-term convergence measurement method based on reflectorless distance measurement technology can be adopted.

2. Detection method for additive constant in reflectorless distance measurement of total station.

The detection of additive constant in reflectorless distance measurement can use direct measurement method or indirect measurement method with test blocks, and the operational requirements have been included in the local standard *Code for Monitoring and Measurement of Urban Rail Transit Structures* in Shanghai.

3.2.2.2 Full Section Scanning and Monitoring Based on Automatic Total Station

Full section scanning and monitoring based on automatic total station is based on Leica's total station with motor driven and infrared ranging at in the late twentieth century, which has solved two technical problems through scientific research:

(1) Research and development of an automated data acquisition airborne program based on total station hardware. The research results include the process and technical requirements for field data collection, achieving standardized and efficient field data collection. Each section automatically collects 80–100 points, including structural feature points such as pipe joints, and on-site collection is shown in Fig. 3.5.
(2) Research on convergence calculation methods and development of post-processing software. The research replaced the "whole section elliptical model least squares method" with the fitting of "multi-arc least squares method", making mathematical model more in line with the deformation characteristics of the lining ring (Wang et al. 2020). The post-processing software achieves automatic gross error removal, data calculation, data warehousing, and report generation by interval, the output results add parameters such as pipe seam dislocation and rotation, and the accuracy information of the data processing process is recorded in a log file. The processing function and automation level are advanced, and the software interface is shown in Fig. 3.6.

Fig. 3.5 Field monitoring of full cross-section scanning with automatic total station

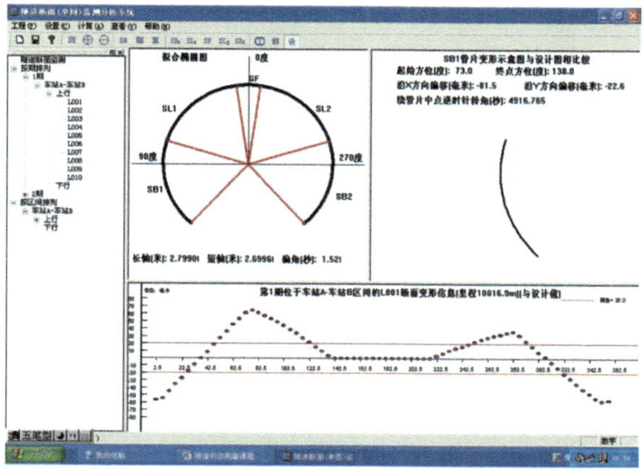

Fig. 3.6 Full cross-section scanning data processing platform

3.2.3 *Structural Disease Inspection*

We have studied and unified disease classification and expression methods, forming the first national urban rail transit "Shield Tunnel Disease Inspection Operation Guide" (Q/SD-WBZ-FB-SS-GW3401-2014), and developed a disease management system and PAD based on-site recording software, as shown in Fig. 3.7.

3.2.4 *Patrol of Protected Areas*

3.2.4.1 Inspection Equipment and System Based on GNSS Intelligent Terminal

In the early 2000s, Shanghai Metro began safety intelligent inspection work. Initially, paper task orders were used for assignment, and external positioning devices and digital cameras were handheld for operations. However, the early inspection equipment was cumbersome, with poor operability, difficult information integration, and difficult quality supervision during inspections. Errors and omissions in inspections were common. Afterwards, the inspection mode of industrial grade handheld terminal equipment + mobile APP operating system + WEB project management and control system based on cloud computing, mobile internet, and GIS system technology was studied. The system platform and equipment are shown in Fig. 3.8. The system has the following characteristics:

(1) With the help of a GIS system supported by mobile networks, data on subway lines and protected areas can be loaded in a hierarchical manner, allowing inspectors to quickly and accurately grasp the location of the subway, ensuring no missed or repeated inspections, and scientifically arranging inspection work.

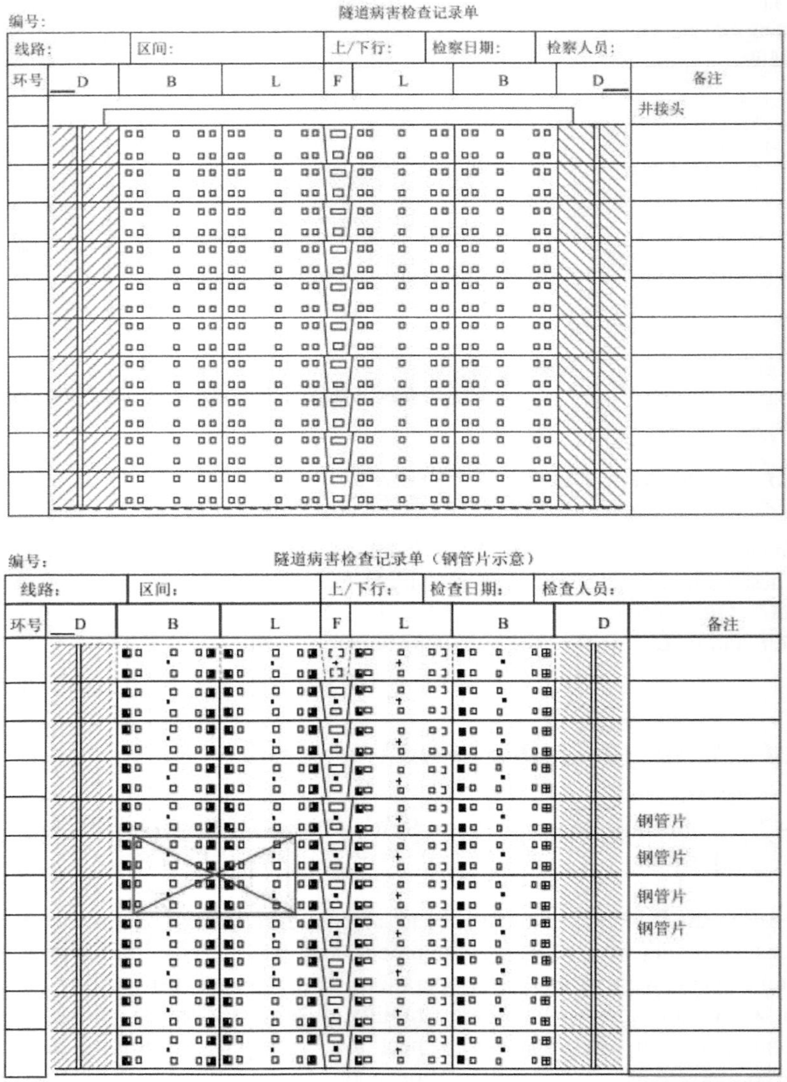

Fig. 3.7 Inspection record sheet for shield tunnel defects

(2) With the help of the GNSS real-time positioning system, the trajectory information of inspectors is continuously recorded, and the function of mapping the affected area of violations is achieved. GIS system archiving ensures the traceability of physical work.

(3) We have developed a fast and efficient on-site collection and release function, utilizing image recognition technology to swiftly gather construction nameplate

Fig. 3.8 GNSS smart terminal for patrol inspection system and equipment

information. Emergency events can be quickly shared through We Chat and have the function of preventing data leakage and scheduled deletion.

(4) A map is used for hierarchical management of protected area projects, achieving full process information management from discovery, application, listing as illegal projects, monitoring projects, and completed projects. It has the functions of report printing and data statistics.
(5) Integrate multi-source data such as geological and structural health monitoring along the rail transit line to assist in the safety management of subway structures.

The product has been in stable operation for four years since it was put into use in the Shanghai Metro, with a cumulative allocation of over 10,000 inspection tasks, achieving a management efficiency improvement of more than 10 times. The product has also been gradually promoted to Tianjin, Dongguan, and other places, achieving good economic and social benefits, providing strong support for the safety management and refined urban management of urban subways in China.

3.2.4.2 UAV Low-Altitude Surveying and Mapping on Protection Area Inspection Technology

Unmanned aerial vehicles (Fig. 3.9) are playing an increasingly important role in the field of low-altitude remote sensing surveying and mapping due to their advantages of high integration, simple operation, fast and efficient operation, flexible maneuverability, and low cost. In recent years, low-altitude remote sensing technology has been used to study the auxiliary inspection technology of rail transit protected areas.

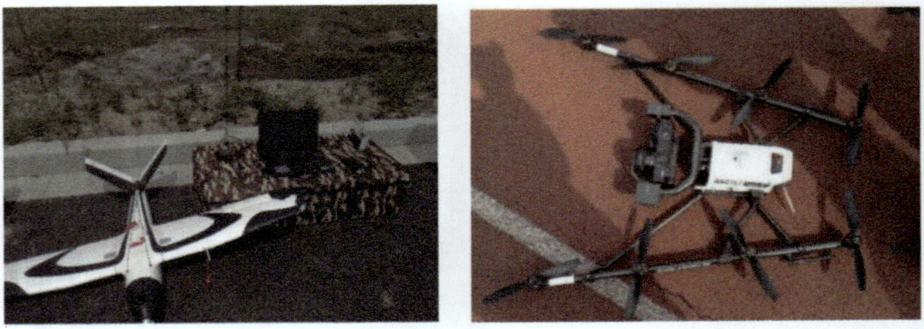

(a)Fixed (Flying Horse F100) (b) Rotor(Falcon8)

Fig. 3.9 Types of drones

The inspection content includes the scope of construction operations or significant changes in terrain and features, the current ground elevation, and the measurement of relative positional relationships. The main product types of drone application achievements include DSM, DOM, and 3D models.

1. Digital Surface Model (DSM)

This is used to check the development of cities, vegetation growth, and construction sites along the rail transit line, it is therefore more suitable for various dynamic monitoring.

2. Digital Elevation Model (DEM)

This is applied to express ground elevation information along rail transit lines, it can be used for quantitative analysis of earthwork, surface elevation, and other aspects in engineering construction.

3. Digital Orthophoto Map (DOM)

DOM can dynamically monitor engineering activities from a macro perspective, and high-resolution images can track engineering conditions and monitor the surrounding environment, as shown in Fig. 3.10.

3.3 Immobile Automatic Monitoring

When there are construction activities such as construction or demolition of buildings (structures), excavation of foundation pits, and underground engineering jacking operations in the vicinity of tunnels, it is necessary to conduct engineering impact monitoring on subway structures. In addition to using general monitoring methods, engineering impact monitoring also requires real-time and online methods of immobile monitoring. Through years of practice and research, the main achievements include:

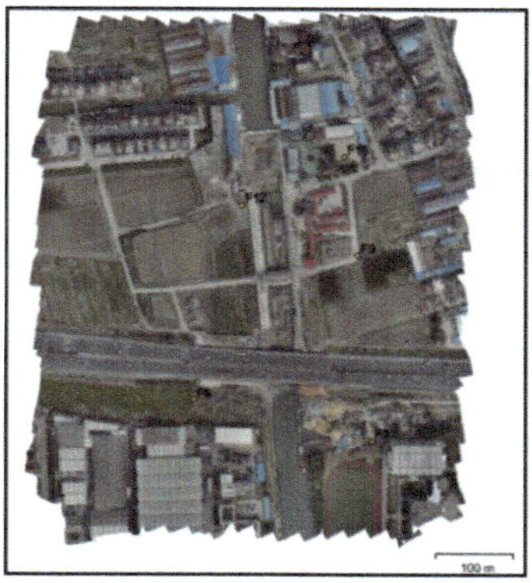

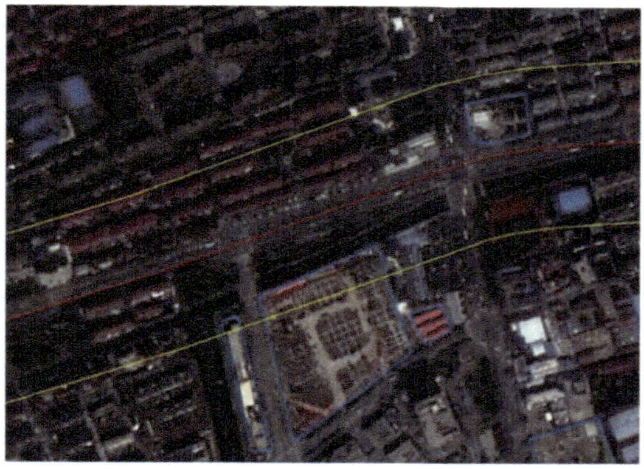

Fig. 3.10 DOM images and application in protected area patrol inspection

(1) We reconstructed the long link and data synchronization technology that connect the front-end sensor to the back-end server, thus achieving remote control order encapsulation and integrated management of multi-source equipment such as laser rangefinder, inclinometer, surveying robot, hydrostatic level, EL beam sensor, crackmeter, and Fiber Bragg Grating based on IPC.
(2) We studied data processing algorithms such as detection of outliers, identification of data exception, and combined adjustment.
(3) We broke bottlenecks of traditional monitoring sensors with multiple types, complex softwares, and data silos, so as to meeting the all-weather remote monitoring needs of key sections of subway tunnels, thus improving the level of intelligent sensing, data coordination management, and early warning.

3.3.1 Reconstruction of Automatic Monitoring System

3.3.1.1 Characteristics of the Internet of Things

The Internet of Things (IoT) collects various information (sound, light, heat, electricity, mechanics, chemistry, biology, location, etc.) in real-time through various devices and technologies such as information sensors, radio frequency identification technology, GPS, infrared sensors, laser scanners, etc. thus achieving ubiquitous connectivity among things-things and things-people, intelligent perception, recognition, and management of items and processes through any possible network connections.

Remote automated monitoring is a typical application of the Internet of Things in the field of structural monitoring, with three main characteristics:

(1) Multi-source perception: Utilizing multi-source sensors such as inclinometers, distance measurement equipment, and surveying robots to obtain information about monitored objects anytime and anywhere.
(2) Reliable transmission: To transmit collected information in real-time and accurate way by integrating various professional networks with the Internet.
(3) Smart processing: To analyze and process massive amounts of data and information by utilizing cloud computing and fuzzy recognition and other technologies, thus monitoring the target intelligently.

The main content of an automated monitoring system includes transaction management, including management of engineering projects, collection terminals, sensors, and definition of calculation methods; and the system collects data regularly according to predefined times, receives measurement commands initiated by users and collects data in real-time, also includes version update, synchronization of configuration, and data and instruction.

3.3.1.2 Framework of Automated Monitoring System

The physical part of an automated monitoring system consists of sensors, collectors (IPC), routers, servers, management centers, etc.

(1) Sensors are terminal devices that directly collect measurement data, which are installed on the structure of the target. The connection between sensors and collectors is divided into two types: wired connection and wireless connection.
(2) The collector (IPC) is installed on the front-end site to control the sensor's collection time, frequency, and temporary storage of data. A collector device can connect sensors of the same or multiple categories, from several to hundreds. The collector synchronizes measurement data and various configurations with the server through the network.
(3) A network router is a connection device between the collector and the server. Typically, a mobile network router can be inserted into the mobile communication network card, which connect to the public network using networks such as LTE (commonly known as 4G). The mobile network router itself can emit Wi-Fi signal for data collector connection, or it can be connected to an IPC with RJ45 network cable. Physical connections such as network cables and optical fibers are also optional for connection.
(4) The server is the hub for managing all integrated smart terminals. On the one hand, it maintains network connectivity with all connected integrated smart terminals through the public network, achieving real-time measurement and data return, additionally, the database on the server stores information such as the configuration, measurement plan, and historical measurement data of the integrated smart terminal, which ensures synchronization with the integrated intelligent terminal; on the other hand, the server is used to forward user operations and terminal messages to the browser.

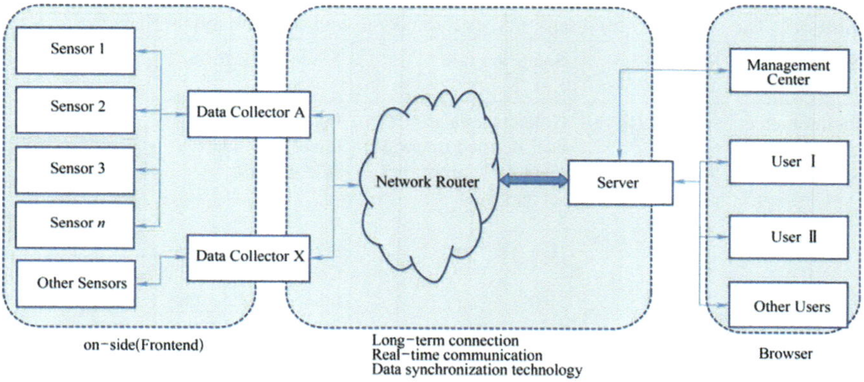

Fig. 3.11 Schematic diagram of automated monitoring system architecture

(5) The management center provides a browser interface for users to control data collection and display, including functions such as sending measurement commands, modifying configurations, and querying measurement data. The logical framework diagram of a typical automated monitoring system is shown in Fig. 3.11.

3.3.1.3 Sensor Types for Underground Structure Monitoring

The commonly used sensor types for underground structure monitoring are shown in Table 3.7.

3.3.2 EL Beam

3.3.2.1 Technical Characteristics

EL beam is an instrument that accurately measures the inclination of objects (which can be converted into the height difference between two points). Multiple EL beams are connected to form a settlement and heave observation system. It was first introduced to Shanghai in 2000 and has been the most widely used automatic settlement monitoring technology for 20 years when previous lines are affected by construction of new lines.

The core part of the EL beam is an electrolyte tilt sensor, which uses electrolytes to measure horizontal deviation (i.e., tilt angle). The structure of an individual unit is shown in Fig. 3.12. An EL beam sensor is made by installing an electrolyte tilt sensor (component) inside a hollow beam. You can install a single or more EL beam sensors in the head-end connection for detection along the direction to be tested within the monitoring section, as shown in Fig. 3.13.

Table 3.7 The commonly used sensor types for underground structure monitoring

Sensor	Physical quantity	Sampling rate	Main application	Actual image
Inclinometer	Angle relative to the horizontal or plumb	Collects data at user-defined intervals, up to 10 samples per second	Inclination of tunnel segments, side wall tilt	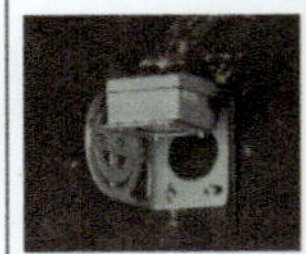
			Combined with horizontal beam to form an electronic level gauge	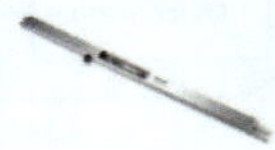
Fixed inclinometer	Angle relative to the horizontal or plumb	Collects data at user-defined intervals, up to 1 sample per minute	Monitoring horizontal displacement in deep soil layers	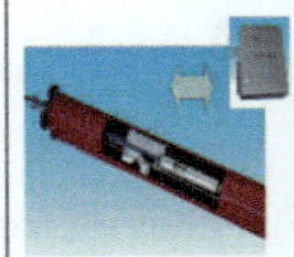
Piezometer	Liquid surface height	Collects data at user-defined intervals, up to 10 samples per second	Structural settlement	
Total station	Horizontal angle, vertical angle, and distance	Collects data at user-defined intervals, typical takes about 5 s for full-circle measurement	3D positioning, displacement, diameter changes	
Distance meter	Distance measurement	Collects data at user-defined intervals, up to 20 samples per second	Tunnel diameter measurement	
Crack meter	Displacement measurement	Collects data at user-defined intervals, up to 10 samples per second	Joint, crack opening measurement	

3.3.2.2 Monitoring Methods

As shown in a monitoring chain, consisting of $E_1, E_2, E_3, \ldots$, and a total of n connected EL beam sensors, starts at measurement point P_0, passes through $P_1, P_2, P_3, \ldots$ and finally reaches P_n. They form a calculation model for monitoring the settlement of

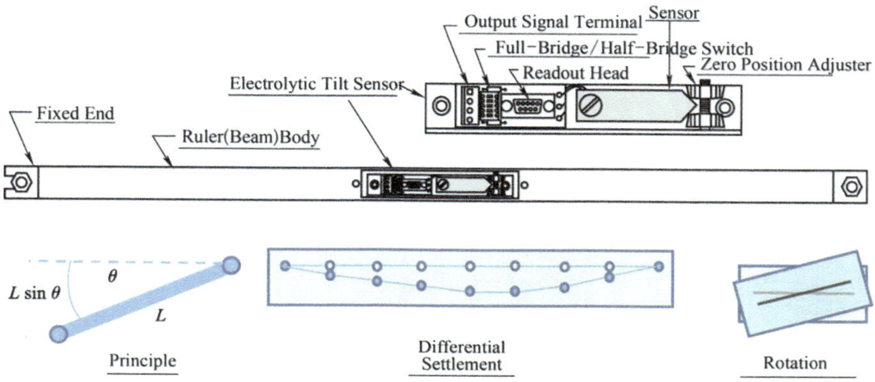

Fig. 3.12 Schematic diagram of the electronic level ruler system structure

(a) Ruler Chain

(b) Scenario for Simultaneous Monitoring of Longitudinal Settlement and Lateral Inclination

(c) Data Collector

Fig. 3.13 On-site installation of an EL beam sensor system

each point within the range of $P_0 \sim P_n$. It has $n + 1$ measurement points, and the length of each EL beam sensor is L_i (usually equal in length) (Fig. 3.14).

The voltage signals output by each sensor are converted into inclination angles, and α_i is the inclination angles of each sensor measured in previous times (i is the measurement point number), and the height difference D_i between each point and

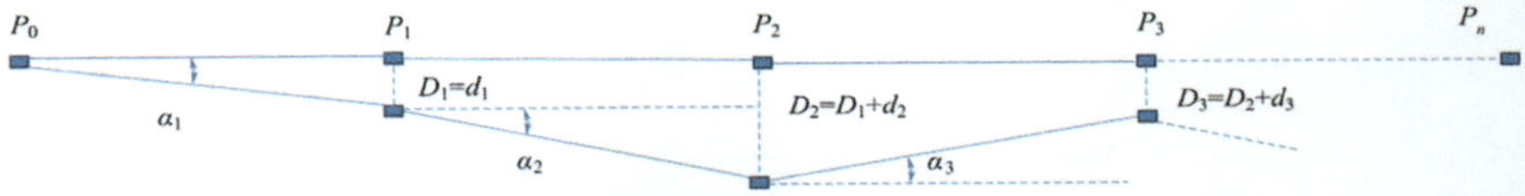

Fig. 3.14 Diagram of settlement calculation using electronic level chain

P_0 point is

$$D_i = \sum_{j=0}^{i} d_j = \sum_{j=0}^{i} \left(L_j \times \sin\alpha_j\right) \tag{3.3.1}$$

Before settlement occurs, we can calculate the height difference D_i^0 between each point and the reference point P_0 (D_i^k for the kth time), and subtract the corresponding values to obtain the settlement of each point ΔD_i^k.

When using a "ruler chain" composed of multiple EL beam sensors for settlement measurement, the leveling method should be used to regularly connect the starting and ending points of the ruler chain, and the settlement deformation values of each measurement point should be corrected based on the leveling results.

3.3.2.3 Precision Analysis and Response Measures

There is a random error in reading of each EL beam sensor, and the accumulation of data from each point on the monitoring scale chain can lead to the superposition of random errors. If the error of a single EL beam sensor is e_i, according to the law of error propagation, the total error of the ruler chain can be estimated as follows:

$$E = \sqrt{\sum (e_i)^2} = e_i\sqrt{n} \tag{3.3.2}$$

Based on experience, the accuracy of the tilt angle change of a single electrolyte tilt sensor is 3″, the accuracy of a single EL beam sensor composed of a 1 m beam is 0.0145 mm (= 1000 mm × 3″/206265″), and the accuracy of that of a 2 m beam is 0.029 mm. The overall accuracy estimation for different types of EL beam sensors to form different length ruler chains is shown in Table 3.8.

In actual monitoring projects, the two ends of the ruler chain, if extending beyond the scope of construction impact, are stable without affection of construction. The

Table 3.8 The overall accuracy for different types of EL beam sensors

Types	50 m chain ruler (mm)	100 m chain ruler (mm)	150 m chain ruler (mm)
1 m EL beam sensor	0.103	0.145	0.178
2 m EL beam sensor	0.145	0.205	0.251

settlement value of the measurement points at both ends should be 0. However, in practice, due to measurement errors, this value is not 0, and the discrepancy value is set to Δ, usually to calibrate by using data adjustment methods

$$H_i = \sum_{j=0}^{i} d_j - \frac{\Delta}{L_T} \sum_{j=0}^{i} L_j \tag{3.3.3}$$

Of course, it should also be noted that not all sites can meet the conditions for stability at both ends of the ruler chain, and regular manual measurement methods should be used to calibrate the measurement points at both ends. Set the initial correction value to H_0 and the final correction value to H_n, and at this point, the discrepancy value Δ, the actual settlement of the detection site i is calculated

$$\Delta = H_0 + \sum_{j=0}^{i} d_j - H_n \tag{3.3.4}$$

$$H_i = H_0 + \sum_{j=0}^{i} d_j - \frac{\Delta}{L_T} \sum_{i=0}^{n} L_j \tag{3.3.5}$$

3.3.3 *Hydrostatic Level*

3.3.3.1 Technical Characteristics

Hydrostatic level is a precision instrument for measuring height difference and its changes. According to different measurement principles, hydrostatic leveling systems are classified into two categories: interconnecting-pipe type and pressure type, as shown in Fig. 3.15.

Interconnecting-pipe measuring system—The liquids in each container are interconnected, liquid exchange exists for measuring the liquid level height.

Pressure measuring system—The connecting pipes between each container are separated by metal membranes, and there is no liquids exchange. The pressure changes at the liquid cross-section are measured.

Hydrostatic level has been applied in the Shanghai subway for over 20 years, and on-site installation is shown in Fig. 3.16.

3.3.3.2 Monitoring Methods

In subway monitoring, interconnecting-pipe measuring system is preferred with its higher stability. In hydrostatic level systems, the vertical displacement of all

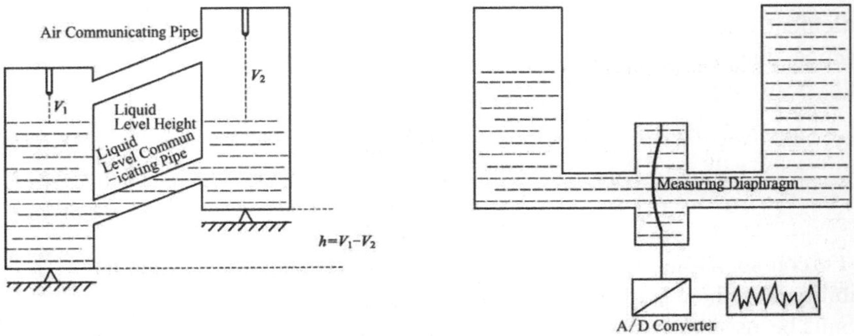

Fig. 3.15 Principle of communicating vessels and differential pressure static level system

Fig. 3.16 Static leveling system installed inside the tunnel

measuring points is based on one point whose vertical displacement is relatively constant or can be accurately determined by other methods, so as to accurately calculate the settlement changes of each measuring point in hydrostatic level systems. The measurement principle of the hydrostatic level is shown in Fig. 3.17.

For line structures with longitudinal slopes, it is often necessary to install survey lines in sections and groups. Two sensors should be set up as transfer points at the intersections of adjacent survey lines above and below the same structure (Fig. 3.18). At the deformation measurement operation site, it is difficult to set the reference points for hydrostatic level to stable areas, and regular geometric leveling joint measurements should be carried out.

3.3.3.3 Precision Analysis

The error influence of the hydrostatic leveling system can be divided into two parts: one part comes from the influence of the external environment, such as uneven expansion of liquid volume and changes in liquid level height under non-uniform temperature and pressure fields; another part of the error comes from the measurement method of liquid level height or pressure difference. For high-precision measurements, various errors should be effectively eliminated or compensated through

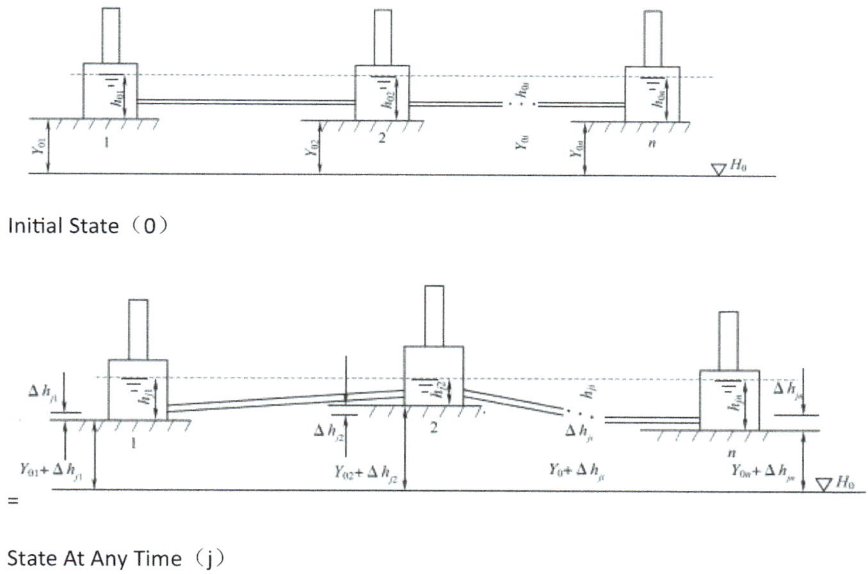

Fig. 3.17 Principle diagram of static level measurement

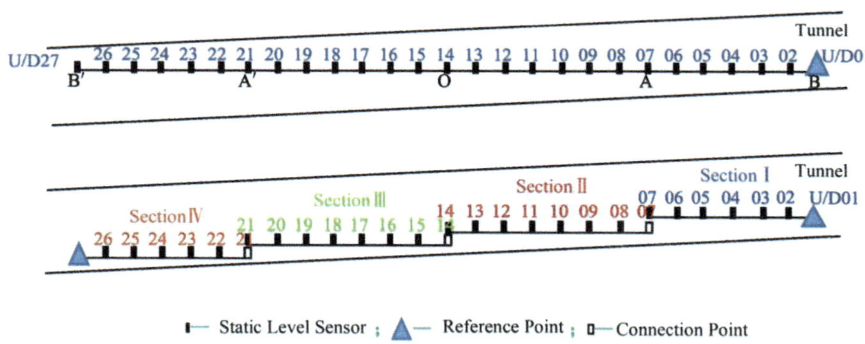

Fig. 3.18 Diagram of static level line segment installation

calculation. The sources of influence and main weakening methods of the external environment are shown in Table 3.9.

Table 3.9 Sources of errors from the external environment and the correction methods

Serial number	Sources of error	Impact modes	Methods of elimination or correction
1	Pressure	Incorrect values due to different pressure levels	Measuring the correct pressure; creating a closed air-pressure system by lengthening the air pipe
2	Temperature	Causing changes in density and volume; different temperatures along the pipeline influencing the accuracy greatly	Remeasuring to obtain the correct temperature; controlling of the height difference from the liquid level to the lowest point of the line; avoiding of direct sunlight or taking measures to keep temperature uniformity
3	Liquid level vibration	Variation around the true value	Remeasuring it until the liquid level is stable or making several measurements over a long period of time and take the average value of the measurements
4	Zero point	Zero-point settlement	Geometric leveling correction
5	Pipeline bubbles	Affecting pressure or liquid level	Precise installation to avoid air bubbles in the pipeline

3.3.4 *Laser Rangefinder*

3.3.4.1 Technical Characteristics

Laser rangefinders have developed rapidly in recent years, with advantages such as fast ranging, small size, and reliable performance. Combined with IoT communication technology, they can automatically monitor in real-time and meet the needs of early warning. They have been widely used in the field of underground engineering. The shield tunnel adopts a laser rangefinder for convergence monitoring, with good sensitivity.

3.3.4.2 Monitoring Methods

Laser rangefinders are typically used to measure changes in the length of fixed measuring lines. The laser rangefinder in the tunnel is usually arranged at a horizontal diameter position, and is usually equipped with a laser rangefinder, a supporting wireless data acquisition module, and a DC 12 V power supply at one end of the measurement line. The attitude of the laser rangefinder measurement line is adjusted to ensure that the direction of the laser rangefinder measurement line is consistent with the design measurement line. After installation, a alignment point is set at the other end to analyze the impact of structural rotation on convergence deformation during operation. During the monitoring process, it is necessary to ensure that there are no obstructions on the measuring line of laser ranging, and regularly use manual

Fig. 3.19 Installation of automatic monitoring point for tunnel diameter measurement using laser distance meter

pipe diameter convergence values to verify automated pipe diameter convergence measurements.

The schematic diagram and example of measuring point embedding are shown in Fig. 3.19.

3.3.4.3 Precision Control

The factors that affect ranging accuracy mainly include the factors of the instrument itself (such as additive constant, multiplicative constant), changes in refractive index caused by atmospheric changes, and the influence of reflective target material.

The deformation monitoring control value of underground engineering is generally 20 mm, and according to the accuracy estimation, it should be within 2 mm. Therefore, a fixed target endpoint should be set, and the accuracy should be controlled from the material, color, transparency, and incidence angle of the ranging laser of the fixed endpoint.

3.3.5 *Inclinometers*

3.3.5.1 Technical Characteristics

Research has shown that the cross-sectional deformation of shield tunnels mainly manifests as dislocation at the segment interface and relative rotation at the joints. By using an inclinometer to monitor the rotation of the segment, the convergence deformation of the tunnel can be analytically calculated. Compared with the laser automatic rangefinder, the inclinometer requires good visibility between the instrument installation position and the opposing aiming point, and the instrument needs to be regularly calibrated after being easily rotated with the attached segment structure; the former does not require visibility and adjustment, with different error sources and relatively complex calculations, making it another solution. the inclinometer

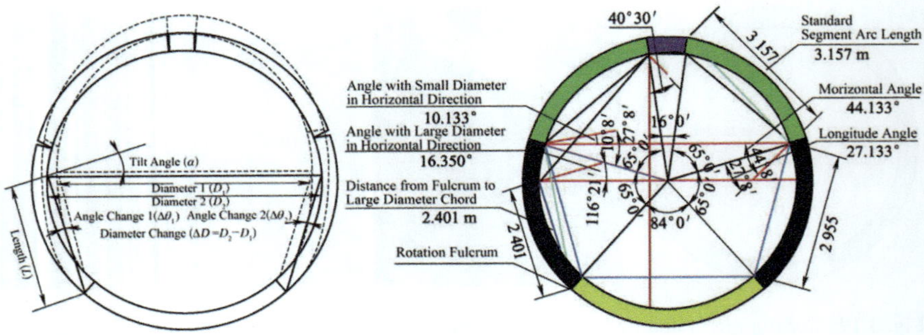

Fig. 3.20 Illustration of the relationship between tilt change and deformation

with wireless micro electro mechanical systems (MEMS) have the advantages of small size, high accuracy in angle measurement, and no wire and no source. They can monitor the deformation of tunnel structures in real-time and achieve monitoring of tunnel convergence under specific algorithms.

The horizontal convergence calculation model for conversion of inclination angle changes is shown in Fig. 3.20. L is the distance from the horizontal point passing through the center inside the intrados of the tunnel to the outer edge of the bottom joint, α is the angle between the vertical line and the horizontal line of the L, D_1 represents the horizontal inner diameter of the tunnel without deformation, and D_2 represents the horizontal inner diameter of the tunnel after deformation, δ_{D} represents the change value of horizontal convergence, which is the difference between D_2 and D_1. The convergence calculation formula is

$$\delta_{\mathrm{D}} = L \times (\Delta\theta_1 - \Delta\theta_2) \times \cos\alpha \tag{3.3.6}$$

where

δ_{D} horizontal convergence variation calculated using the inclination sensing method.

$\Delta\theta_1 - \Delta\theta_2$ inclination variation values at the two mounting points on the standard block, respectively.

3.3.5.2 Monitoring Methods

During operation, a pair of monitoring sensors are generally installed in the standard block and adjacent block for verifying and comparing with each other, as shown in Fig. 3.21.

The convergence changes from the change and conversion of inclination angle installed on a section of a certain grouting reinforcement project are shown in Figs. 3.22 and 3.23.

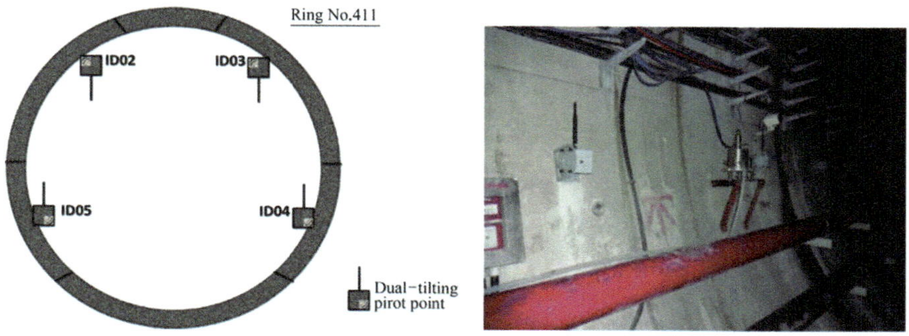

Fig. 3.21 Installation diagram for convergence measurement of inclination sensor

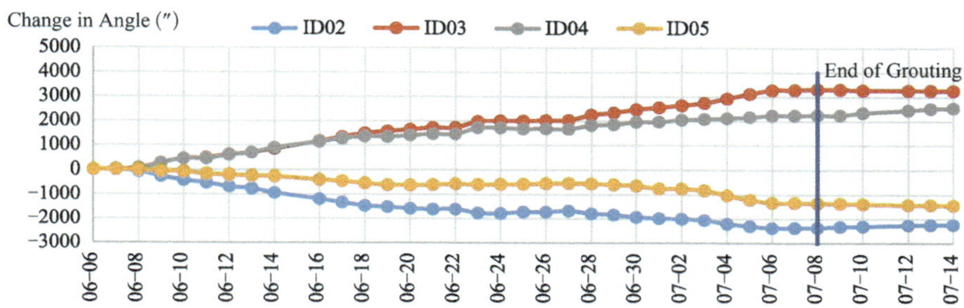

Fig. 3.22 Time series curve of tilt angle changes

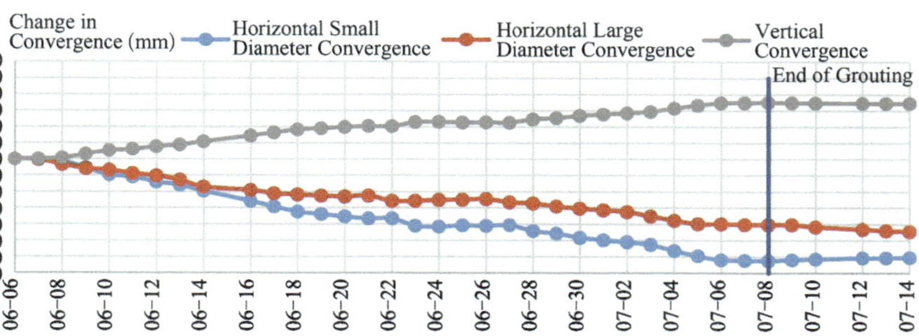

Fig. 3.23 Time series curve of convergence changes

3.3.5.3 Precision Control

There are two main sources of error when using inclinometers for convergence measurement:

(1) In terms of the model, the actual geometric characteristics of the structure are not completely consistent with the model, and the angle relationship after deformation is not rigorous; at the same time, the convergence changes caused by dislocation were not considered in the calculation.
(2) The impact of inclination measurement accuracy. After extensive monitoring and verification, the comprehensive accuracy impact is not more than 10%.

3.3.6 Comprehensive Monitoring of Georobot

3.3.6.1 Technical Characteristics

GeoRobot is an intelligent electronic total station that can automatically search, identify, and accurately target, and automatically obtain measurement information such as distance, angle, three-dimensional coordinates, and images. GeoRobot provides users with an open data interface, making it possible to support automatic monitoring of comprehensive deformation.

Compared with the EL beam sensor system and hydrostatic leveling system, Georobot comprehensive monitoring system has the following characteristics:

(1) The measurement range is long, and the setting range of datum point is large, making it easy to obtain the "absolute" deformation of the relatively stable reference point. It is suitable for situations with a large range, uniform accuracy requirements, and moderate level (1–2 mm level). However, the cost of extending the coverage range of systems such as EL beam sensor is high, and the datum points often require other means of regular detection, which is more suitable for high relative accuracy requirements (sub-millimeter level) in thousands of parts.
(2) Georobot essentially measure the three-dimensional coordinates of spatial points, thus giving us comprehensive results after calculating multi-type deformations, such as settlement, displacement, and convergence, through the comparison of coordinate components; compared to other monitoring systems that can only monitor one measurement item, one physical or quantitative quantity, the results are more diverse in content.
(3) The cost composition is also different, with high prices for measuring robots and low prices for target points, making it suitable for situations with multiple monitoring points and dispersed locations, in other word, high extensibility; other monitoring systems are generally composed of multiple essentially equivalent sensors in tandem, with stricter requirements for height or linearity during on-site installation.
(4) In terms of monitoring speed, the Georobot adopts a loop-monitoring to aim and measure one by one, and each monitoring point takes several seconds, which is slow and time-consuming; other sensors use some form of physical induction, resulting in relatively faster data collection.

3.3.6.2 Monitoring Methods

Automatic measurement of Georobot is suitable for three-dimensional coordinate measurement. The Georobot and the observation target set on the monitoring object are combined to form an observation system. Based on the different corresponding relationships of the observation points, measurement results such as horizontal displacement, pipe diameter convergence, settlement, or structural inclination can be obtained.

1. Arrangement and Installation of Measuring Points

   (1) Georobot setup: Fixed instrument observation stations are generally installed on-site, with a mandatory centering device installed on the top of the observation station. The total station is installed on the mandatory centering device and connected to the data collector through communication cables. Fixed instrument observation stations are installed on the inner wall of the shield tunnel structure, station side walls, or within the tunnel or station (meeting the requirements of rail transit clearance). The form of observation stations generally includes instrument brackets or metal tripods. The instrument brackets are drilled and installed on the structure through expansion screws, and the metal tripod base is fixed by inserting the metal angle iron base into the structure through expansion screws, as shown in Fig. 3.24.

Fig. 3.24 Installation photos of total station

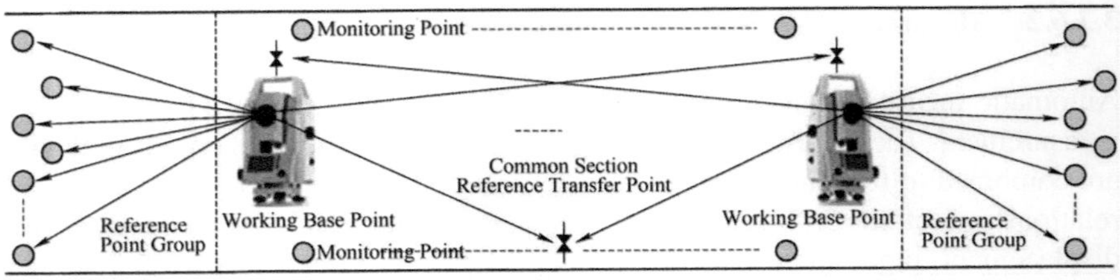

Fig. 3.25 Multi-station data processing model diagram

(2) Cooperation goal setting: The cooperation goal is usually a small prism, which is usually set after the on-site instrument installation. The visibility between the two should be verified through the instrument, and the installation should avoid mutual interference of multi prism in the same direction of the instrument. The staggered angle between cooperation targets should meet the requirement of angle of minimum resolution.

The self-colored highlight in the right figure in Fig. 3.24 is the prism monitoring point.

2. Data Processing

Compared to the single grouping and simple calculation of measuring points such as hydrostatic level and EL beam sensor, the data processing program of Georobot is more complex. During the calculation, the observation tolerance check, measurement deviation check, and control point stability check of field results should be carried out. After the calculation, accuracy statistical information should be output.

A set of data processing algorithms should be specifically developed on the server for data processing, typically with the following functions:

(1) Support one or more total stations, which can adapt to a wide range of monitoring intervals.
(2) Multiple total stations can work together to start measuring at the same time, ensuring the timeliness of data.
(3) Using an indirect adjustment model, the total station can set up stations in the deformation area. After the observation is completed, the overall adjustment automatically corrects the coordinate changes of the stations. The schematic diagram of the data processing model is shown in Fig. 3.25.
(4) According to the corresponding relationships of different measurement points, various measurement results can be automatically calculated, such as pipe diameter convergence, roadbed settlement, horizontal displacement, or side wall inclination.

3.3.6.3 Precision Analysis

The accuracy of the observation results of measuring robots is mainly influenced by error sources such as benchmark stability, instrument timing angle measurement, and distance measurement accuracy. There are many influencing factors, which can be seen in relevant content such as "measurement adjustment".

3.3.7 *Fiber Bragg Grating Stress Strain Monitoring*

3.3.7.1 Technical Characteristics

Fiber Bragg Grating not only has the advantages of fiber optic sensors, but also has the advantages of strong wavelength separation ability, insensitivity to environmental interference, extremely high sensing accuracy and sensitivity, and convenience for absolute digital measurement and precise positioning. Especially in recent years, the dense distributed fiber optic grating sensors developed based on time division multiplexing technology can achieve hundreds or thousands of grating sensors connected in series with a single fiber, changing the limitations of traditional distributed fiber optic grating point monitoring and greatly improving the load-bearing capacity of a single fiber; further integrating the optical fibers of multiple Fiber Bragg Grating sensors into a spatially distributed sensing network system, which is considered a promising sensing method currently.

3.3.7.2 Integrated Technology of Fiber Bragg Grating Sensing System

1. Communication of Fiber Bragg Grating Monitoring System

The connection methods within the cross-section of fiber optic sensors are divided into parallel and series types. In practical engineering, to ensure the survival rate of sensors, it is not advisable to connect all sensors in series. Figure 3.26 shows the topology structure of the cross-sectional sensor link, and Fig. 3.27 shows the layout of on-site sensors.

The connection of cross-section sensors is shown in Fig. 3.28. Sensors in each channel are connected into the main optical cable via a fiber bundle combiner. According to the structure of the optical cable and the transmission characteristics of light in the optical fiber, each fiber in the main optical cable can serve as a

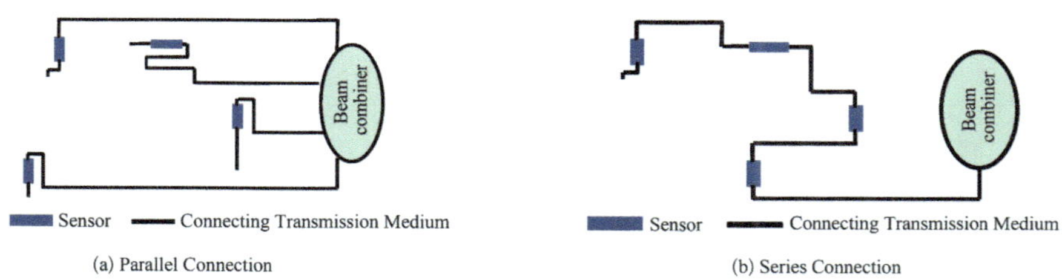

Fig. 3.26 Connection layout of sensors in cross-section

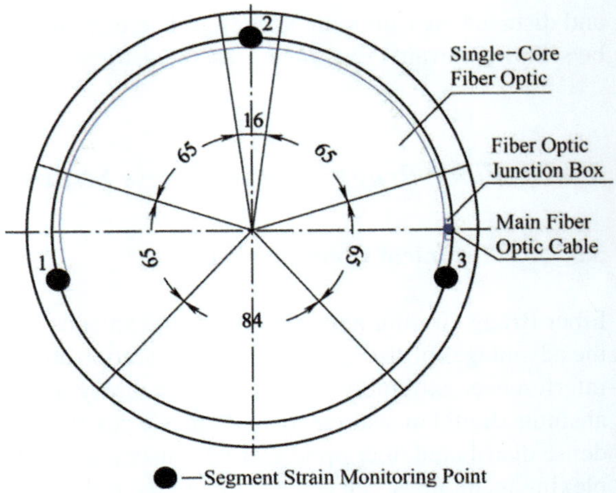

Fig. 3.27 Diagram of cross-section layout for strain measurement

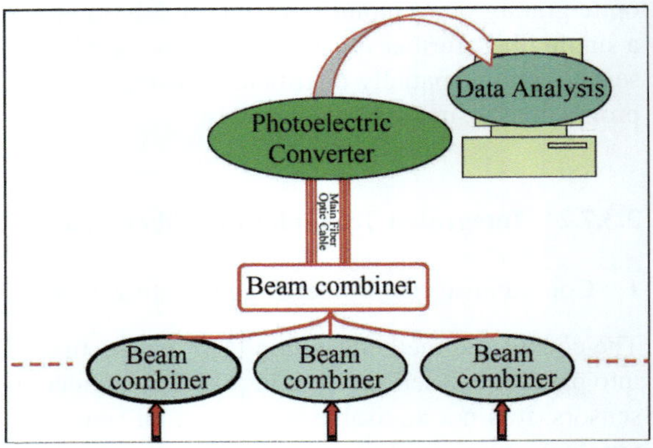

Fig. 3.28 Diagram of cross-section connection for fiber optic sensors

signal transmission channel. Data enters the signal demodulator through the main optical cable and is converted into digital information recognizable by the host via an optoelectronic converter. Within the Fiber Bragg Grating sensing network, channels may be connected either in series or parallel.

2. Fiber Bragg Grating Monitoring System Model

The modern tunnel safety prediction and early warning system should be a software and hardware integrated system that integrates modern advanced technology, adapts to people's efficient and safe traffic concepts, evaluates tunnel traffic capacity, monitors the normal operation of the tunnel, warns of safety hazards in the tunnel structure, assists in handling tunnel traffic emergencies, and ensures the normal functioning of the tunnel. The specific functions are: monitoring of vehicles passing through the

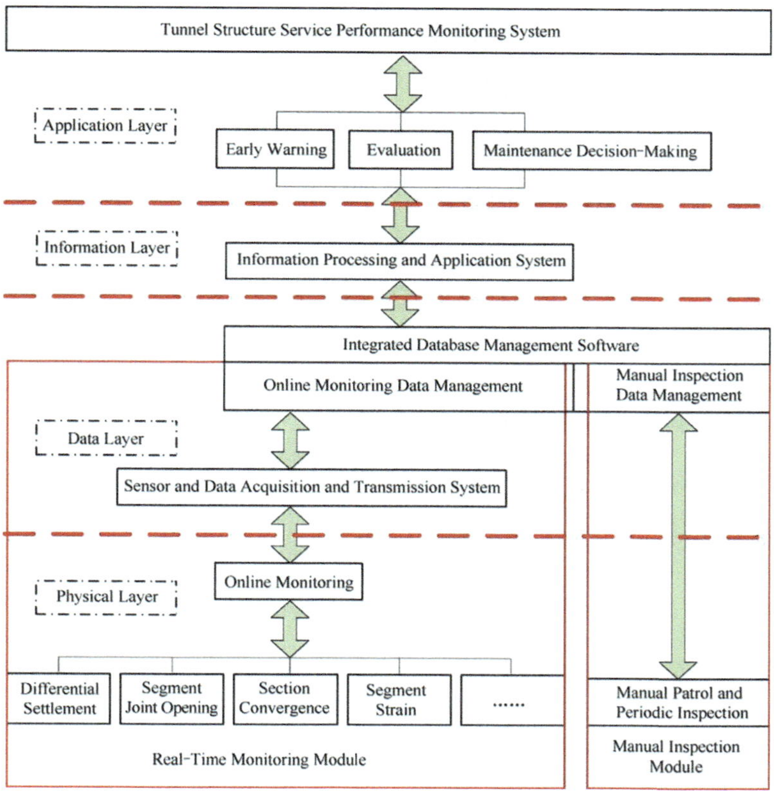

Fig. 3.29 Tunnel structure service performance monitoring system

tunnel and emergency events; health monitoring function for the occurrence conditions of tunnel structural systems; real time monitoring function for tunnel operating environment; the safety status evaluation function of tunnel structure system; tunnel capacity evaluation and prediction function; tunnel traffic safety identification and decision-making function; system integration and control functions. The tunnel structure service performance monitoring system is shown in Fig. 3.29.

3.4 Mobile Detection

Immobile automatic monitoring yields highly accurate results but has few measuring points, limited coverage, and requires a substantial maintenance workload. As the total tunnel mileage and subway operation times increase, the workload for tunnel structural disease inspection grows, and the available time window for inspections shortens, making it difficult for traditional manual inspections to meet actual needs.

Through years of continuous research and development, significant advances have been made in laser scanning detection, visual disease detection, and structural non-destructive testing, which have been applied in practice.

3.4.1 Comprehensive Deformation Detection of Tunnel Structures Based on Laser Scanning

Laser scanning technology, also known as "real scene replication technology", breaks through the basic process of traditional measurement from aiming to measuring then recording. It can accurately and efficiently obtain the three-dimensional coordinates and image data of the measurement target in a full field of view. It has advantages such as high measurement efficiency (1 million points/second), rich measurement information (coordinates + laser reflectivity), and high measurement accuracy (mm level). The application of laser scanning in tunnel detection can achieve two results: on the one hand, the geometric information of point clouds collected by laser scanning can be used to solve structural convergence, dislocation, and other deformations; on the other hand, the use of point cloud reflectance can generate images of the inner walls of tunnels, thereby meeting the requirements of disease inspection.

Through years of research, the laser scanning detection technology for shield tunnel diseases has been achieved, and the operational efficiency has been improved by more than 50–100 times. The main achievements include:

(1) Designing and producing two tunnel mobile scanning and inspection vehicles, achieving technical indicators such as autonomous power, high-precision uniform speed, and low vibration. The equipment integration has been greatly improved and the cost has been greatly reduced compared to foreign countries.
(2) Studying the selection of operation methods, parameter settings, and optimization combinations for tunnel laser scanning, and to prepare a set of operation guidance documents, and propose a set of on-site operation and quality control indicators.
(3) Researching tunnels with different lining forms such as single circles, large circles, and rectangles, and proposing core algorithms and processing processes such as image distortion correction, asymmetric lining structural deformation, disease thematic mapping, and BIM real scene modeling.
(4) Developing a set of data collection and processing software to standardize the entire process from on-site data collection to in-house processing, so as to reduce intermediate processes and operational settings, and break the monopoly of foreign software products.
(5) Based on the cloud platform, the SAAS service model for data processing and result hosting is implemented, which improves data collection efficiency through parallel computing and greatly reduces the threshold for software deployment, use, and promotion.

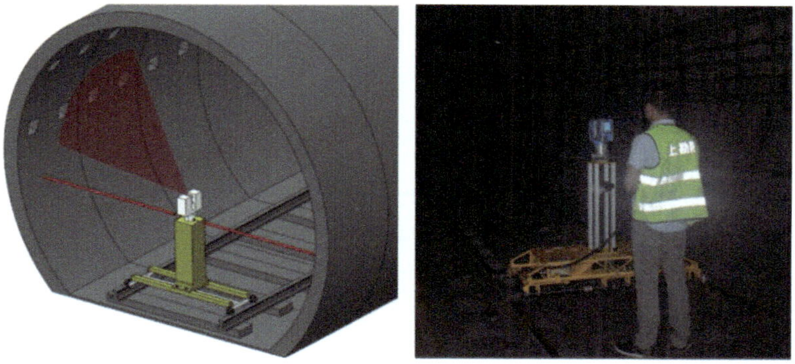

Fig. 3.30 Principles and actual image of tunnel mobile scanning measurement vehicle

Over 1000 km of engineering practice have verified the measurement accuracy and quality reliability of mobile laser scanning technology, with a deformation measurement accuracy of ± 3 mm, a maximum image resolution of 3 mm, and an average operational speed of 3 km/h.

3.4.1.1 Data Collection

Mobile laser scanning technology utilizes a specialized mobile platform, equipped with a laser scanner for cross-sectional scanning to obtain continuous geometric coordinates and laser reflectivity information within the tunnel. It accurately records the inspection vehicle's mileage information in sync and calculates structural deformations and inner wall laser images through software, achieving quantitative and qualitative assessment of tunnel diseases. Its working principle and the actual equipment are shown in Fig. 3.30, with the mobile inspection vehicle's operational speed adjustable between 0.9 and 5.4 km/h based on resolution requirements.

3.4.1.2 Data Analysis and Calculation

1. Data Processing

The processing content of scanned data in the office mainly includes: (1) Initial image calculation, mainly including time system matching of mileage data and point cloud data, initial image calculation, etc.; (2) Tunnel section ring number positioning; (3) Section slicing extraction and horizontal diameter calculation; (4) Correction of tunnel inner wall images. The specific steps of data processing are shown in Fig. 3.31.

2. Tunnel Point Cloud Data Denoising

There are inevitably many noise points in the massive point cloud data obtained from laser scanning of the final image correction, which can be roughly divided

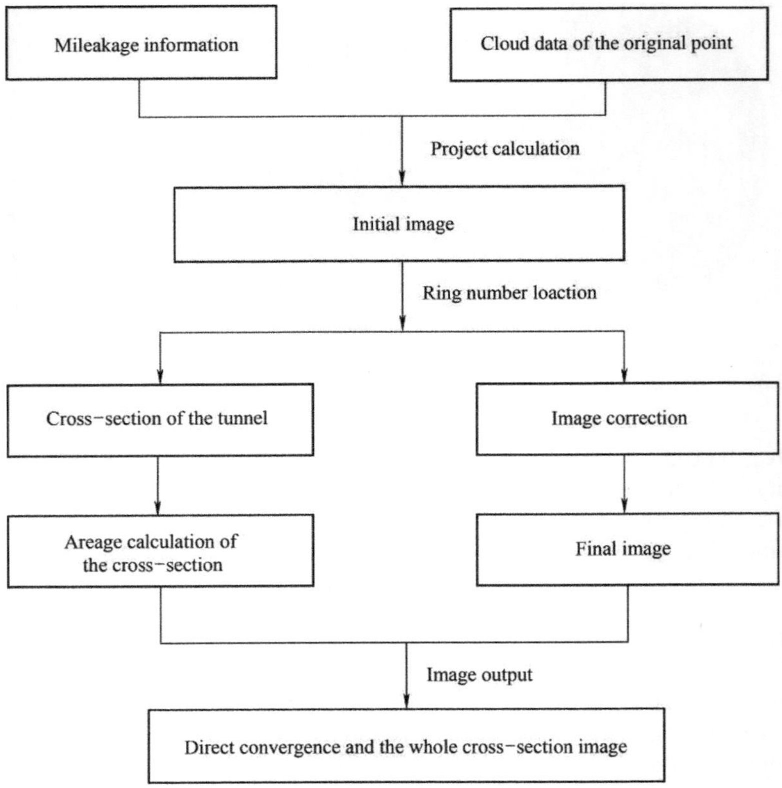

Fig. 3.31 Mobile scanning data processing flow chart

into drift points, redundant points, and mixed points. In tunnel data processing, drift point noise are independent of the tunnel and far away from the main part of the tunnel. It is usually removed directly through existing point cloud processing software. Redundant noise points and mixed noise points are related to the tunnel and mixed with the main point cloud of the tunnel, as shown in Fig. 3.32. The denoising algorithm mainly deals with noise such as redundant points and mixed points unrelated to the tunnel structure.

3. Calculation of Horizontal Diameter of Lining Ring

A large amount of observation data indicates that tunnel deformation is mainly reflected in the cracks in the splicing of pipe segments, and the cross-section deformation is not entirely a standard ellipse. The method of segmented arc fitting is closer to the real situation when fitting the point cloud of tunnel cross-section. The segmented arc fitting method for tunnels mainly uses the center, radius, and starting

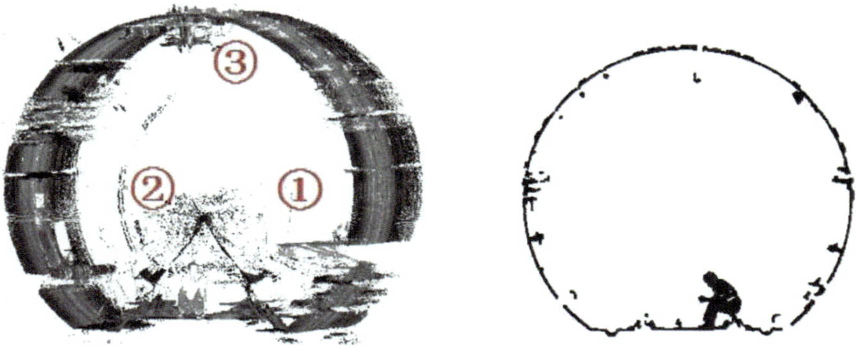

Fig. 3.32 Tunnel noise distribution map

Fig. 3.33 Segmented arc fitting schema

and ending angles to describe the spatial positions and interrelationships of each fitted arc, as shown in Fig. 3.33.

Due to the fixed relative position of the joints of each arc in each section of the tunnel, the α (starting angle) and β (ending angle) of the arc can be preset according to the actual situation. Therefore, the tunnel circle equation can be written as

$$F = \begin{cases} (x - X_1)^2 + (y - Y_1)^2 = R_1^2 \alpha_1 < \theta < \alpha_2 \\ (x - X_2)^2 + (y - Y_2)^2 = R_2^2 \alpha_2 < \theta < \alpha_3 \\ (x - X_3)^2 + (y - Y_3)^2 = R_3^2 \alpha_3 < \theta < \alpha_4 \\ (x - X_4)^2 + (y - Y_4)^2 = R_4^2 \alpha_4 < \theta < \alpha_5 \\ (x - X_5)^2 + (y - Y_5)^2 = R_5^2 \alpha_5 < \theta < \alpha_6 \end{cases} \tag{3.4.1}$$

$$\theta = \arctan \frac{x - X_0}{y - y_0} \tag{3.4.2}$$

The distance d from a point on the circle to the fitting center $C(X_0, Y_0)$ of the tunnel circle can be written as

$$d = \sqrt{(x - X_0)^2 + (y - Y_0)^2} \tag{3.4.3}$$

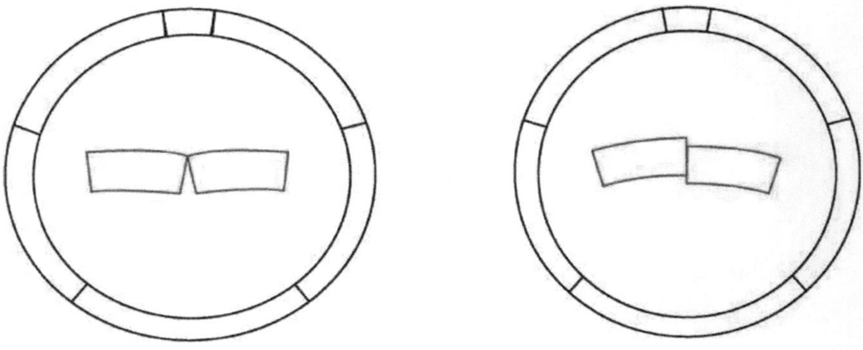

Fig. 3.34 Relative rotation and dislocation of tube segments

4. Extraction of Lining Ring Dislocation and Joint Opening

The dislocation of shield tunnel refers to the phenomenon of adjacent segment walls not in a curved surface after assembly, as shown in Fig. 3.34.

Setting the zenith distance between adjacent arc joints as α, then calculating the coordinates (x_1, y_1) and (x_2, y_2) of two circular arcs at the joint. The value of dislocation is calculated by

$$d = \left|\sqrt{(x_1 - x_0)^2 + (y_1 - y_0)^2} - \sqrt{(x_2 - x_0)^2 + (y_2 - y_0)^2}\right|$$

(x_0, y_0) is the center of the cross-section.

5. Image Generation of Segmented Cylinder Using Isometric Tangent Azimuth Projection

The point cloud data of each ring can be abstracted as a cylindrical model, then to expand side views of each cylinder into a plane according to mathematical projection principles, with keeping their continuity along the mileage direction, so as to generate an orthophoto projection of the tunnel cross-section. The mobile scan is approximately perpendicular to the track. Due to the continuous and smooth nature of the track curve, it does not require axis rotation or slope correction after projection, and can be directly concatenated to form a two-dimensional plane of the tunnel, as shown in Fig. 3.35.

3.4.1.3 Accuracy Evaluation

1. Precision of Inner Coincidence Evaluation

Same instrument and measurement method would be performed on the same subway with for twice, so as to evaluate the accuracy of the 3D scanning data. After analyzing

Fig. 3.35 Raster image of tunnel wall

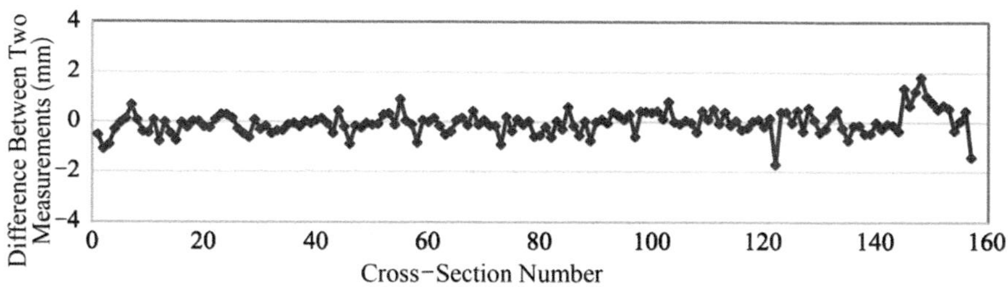

Fig. 3.36 Repeated measurement difference curve of laser scanning for a section

Table 3.10 Statistics of laser scanning repeat measurement differences

Range (mm)	Number	Percentage (%)
$-\sigma \sim \sigma$	7669	81.0
$-2\sigma \sim 2\sigma$	9300	98.3
$-3\sigma \sim 3\sigma$	9449	99.9

9463 sets of repeated data, the comparison of repeated measurement data in a certain interval is shown in Fig. 3.36.

Statistical analysis was conducted on the difference between the 9463 sets of repeated data mentioned above, with a maximum value of 2.7 mm, a minimum value of − 2.8 mm, an average value of − 0.1 mm, and a mean square error σ 0.6 mm. The detailed statistical results are shown in Table 3.10.

2. Precision of External Coincidence Evaluation

Table 3.11 Laser scanning external conformity accuracy statistics

Range (mm)	Number	Percentage (%)
$-\sigma \sim \sigma$	1093	69.5
$-2\sigma \sim 2\sigma$	1510	96.1
$-3\sigma \sim 3\sigma$	1558	99.1

Two measurement methods, total station and scanner, are used for convergence measurement of the same tunnel section, and the accuracy of the scanner data is evaluated based on the total station data. After statistical analysis of 1572 sets of measurement data, it was found that the maximum value was 7.5 mm, the minimum value was − 8.4 mm, and the average value was − 0.3 mm, with a mean square error σ. It is 2.2 mm, and after field verification, data larger than 5 mm are all caused by the inconsistent placement of points in traditional detection methods and laser scanning calculation positions. The detailed statistical results are shown in Table 3.11.

3.4.2 Rapid Detection of Tunnel Structural Diseases Based on Camera Group

A mobile comprehensive detection equipment for the safety status of subway tunnel structures based on high-speed cameras can be used for on-site collection, extraction, identification, and analysis of tunnel surface diseases. The developed system includes an information collection system for tunnel surface diseases based on industrial cameras and professional lighting distribution design, as well as a supporting post-processing software for identifying surface diseases (such as structural leakage, concrete damage) based on the differences in image characteristics of tunnel structural segment lining. An image processing-based tunnel surface disease detection method is proposed. Based on the principle of photogrammetry, the system has developed a vehicle mounted measurement device and related system for quickly obtaining tunnel cross-section data during movement by taking a structured light source in conjunction with an industrial fixed focus camera (Specialized rail car equipped with tunnel cross-section scanning system (ZI201020292738.7)).

This equipment has the characteristics of module integration, convenient assembly, compact structure, fast detection, automatic processing, and statistical analysis. The on-site testing speed of the equipment reaches 5 km/h, and the efficiency is more than 200 times higher than manual detection, the capture and recognition rate for dangerous structures deformation can reach as high as 100%, and for the major apparent disease the capture and recognition rate is above 95%. The specific workflow of automated collection of apparent diseases is shown in Fig. 3.37.

Key equipment technologies include positioning technology, data processing technology, identification algorithm for surface disease of tunnel structure.

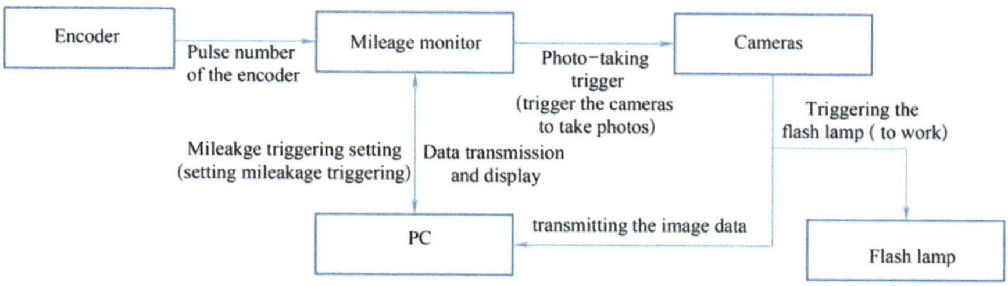

Fig. 3.37 Flow chart of automatic collection of apparent tunnel damage

3.4.2.1 Positioning Technology for Surface Disease of Tunnel Structure

The positioning technology mainly relies on encoders and mileage controllers. The encoder can record the location of the mileage, and the mileage controller adds this information to the image data to achieve precise positioning of the disease. The encoder outputs a fixed number of pulses evenly and stably every revolution, and the circumference of the wheel is a fixed value. By fixing the encoder and the wheel together, the mileage position of the image can be calculated by recording the total number of pulses from the encoder. If the encoder outputs a pulse every revolution and the circumference of the wheel is C, and the encoder has already emitted a pulse when capturing the image, then the mileage position M where the image is located is.

In addition, the apparent disease collection system may also encounter back-tracking during the travel process. In order to ensure the accuracy of mileage calculation, the mileage controller has designed a set of algorithms based on the characteristics of the encoder's output pulse signal.

When the encoder is working, it outputs two signals, A and B, and the B signal always lags behind the A signal by $\pi/4$ wavelengths. After receiving two signals from the encoder, the mileage controller determines whether the apparent disease collection system is moving forward or backward based on whether the signal lags behind the A signal $\pi/4$ wavelength (the signal always lags behind the A signal $\pi/4$ wavelength when moving forward). If the mileage controller determines that the apparent disease collection system is moving forward, the number of pulses received is added to the original number of pulses. If backward, subtract the number of received pulses from the original number of pulses. After the mileage controller receives two signals from the encoder, this calculation method can eliminate the errors caused by the calculation when the apparent disease collection system brakes and retreats based on whether the signal is behind. It ensures that each photo taken by the apparent disease collection system must correspond to the actual mileage in the photo, avoiding repeated shooting, overlapping shooting, incomplete shooting information, and other factors that may affect the detection effect.

3.4.2.2 Data Processing Technology for Surface Disease of Tunnel Structure

The system consists of four cameras, each covering a certain angle range, and ultimately capturing images of the inner surface of the entire tunnel. In order to collect data simply and normatively, the data is divided into four channels according to the camera number, and each channel ultimately stores only one *. avi file, which can be viewed using universal image playback software. The system processes the data collected from images as follows:

(1) Creating a stream file corresponding to the channel, and file is named after the channel number.
(2) Number the image through the trigger count value of the mileage controller, and place the image in the corresponding frame number position.
(3) When closing the collection, the stream file is closed to form a channel number *. avi file.

In practical application, the corresponding image data is extracted from the *. avi file using the channel number and frame number for analysis. This not only simplifies the operation of the file, but also effectively reduces the size of the file.

3.4.2.3 Identification Algorithm for Surface Disease of Tunnel Structure

The automatic recognition of surface diseases mainly relies on image processing technology. Due to the differences in shape, color, and other characteristics between surface diseases such as water leakage, damage and healthy segments, image processing technology can be used to identify, extract, and calculate the characteristics of diseases. If the above characteristics of a certain area are found to be different from healthy segments, after analysis and confirmation, the area can be marked as a disease.

The system recognizes, extracts, and calculates diseases through three main steps: image preprocessing, image segmentation, disease feature calculation, and disease classification.

(1) Image Preprocessing

The main purpose of image preprocessing is to filter out noise in the image, so as to enhance the image features of diseases, improve the extractability of diseases, and facilitate subsequent image segmentation and feature calculation. The image preprocessing techniques used in this system mainly include grayscale and image filtering.

(2) Image Segmentation

Image segmentation is to separate the background and disease for extracting a complete tunnel disease image, then preparing for the next step of feature calculation. The erosion algorithm and dilation algorithm are also used in this program, using an open operation of erosion before dilation. Erosion would help to avoid mistakenly identifying noise as a disease by reducing noise in images; dilation can fill voids and edge burrs caused by segmentation, improving the accuracy of disease calculation features.

For the processing of binary images, the intuitive explanation of dilation is: after mapping structural element B, it moves on image A. When the images of A and B intersect, the set of all points passing through the origin of B's image is the result of B dilation A. The specific principle is shown in Fig. 3.38. The intuitive explanation for erosion is that when structural element B is completely contained in image A, the set of origin positions of B is the result of B corroding A, as shown in Fig. 3.39.

The binary images generated after threshold segmentation mainly include pockmarks, holes, and broken disease areas. The effect of erosion is to remove excess pits and edge burrs from the image, reducing the recognition error rate. The result of dilation is to connect the holes in a certain diseased area and connect two adjacent diseased areas, reducing the number of diseased areas.

(3) Calculation of Disease Characteristics

The calculation of disease characteristics mainly relies on the connected area algorithm, which can obtain feature information such as area size, duty cycle, aspect ratio, and grayscale difference.

(4) Disease Classification

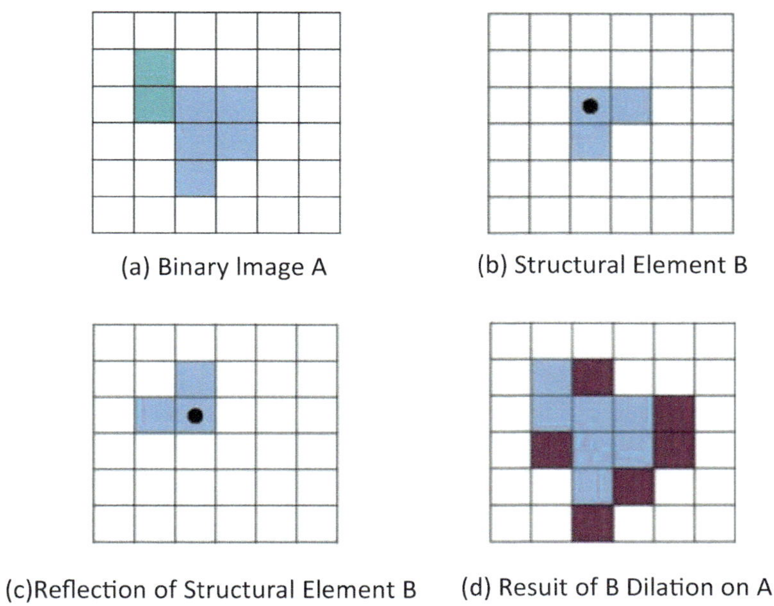

Fig. 3.38 Results of the binary expansion operation (the darker part is the enlarged one according to the original image)

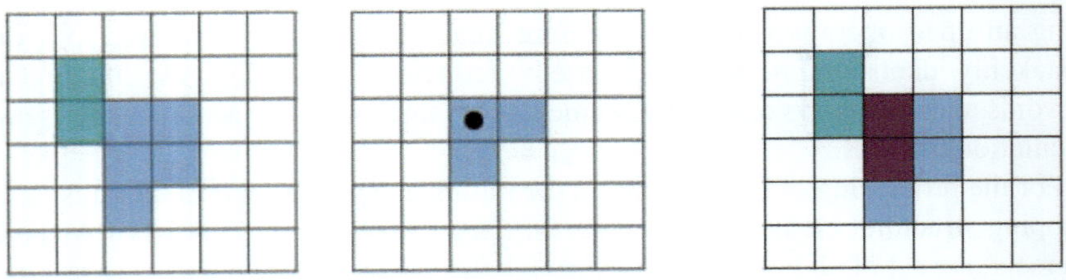

Fig. 3.39 The result of the binary erosion operation (the darker part is what's left compared to the original image)

Table 3.12 Characteristics of connectivity zones for different diseases

Type of diseases	Area size	Spatial ratio	Length–width ratio	Grey level difference
Cracks	Small	Small	Oldest	Small
Water seepage	Large	Large	Moderate	Oldest
Damage	Moderate	Large	Small	Moderate

To classify diseases by analyzing their characteristic information, such as area size, duty cycle, aspect ratio, and grayscale difference. There are some statistical rules in the characteristics of diseases such as water leakage, damage, and cracks. Through continuous learning and correction, the program can obtain appropriate parameters to classify diseases.

Based on the experience of manual detection and recognition in tunnels, the characteristics of water leakage, damage, and cracks vary, but the characteristics of the same disease have certain patterns. Table 3.12 shows the characteristic information of various diseases.

Based on the different feature information of various diseases mentioned above, the size threshold can be determined through statistical experience to identify the disease type to which each disease connected area belongs. In addition, the position of the disease can be located based on the mileage recorded by the detection system for each photo. The spatial distribution of diseases within the tunnel can be calculated based on their size, type, and location.

3.5 Summary

After more than 20 years of maintenance and support practice in Shanghai Metro, a series of mobile rapid comprehensive detection equipment has been developed to achieve synchronous automatic collection and real-time warning of multiple tunnel diseases; we have created a multi-source sensing integration platform to enhance the level of intelligent perception, data coordination management, and early warning; we have established a monitoring system that can comprehensively perceive changes in the tunnel laying environment, improving our ability to control external environmental impacts, mainly reflected in the following aspects:

(1) A monitoring benchmark framework with 45 sets of bedrock markers has been established, covering Shanghai. We've also developed inspection equipment based on GNSS smart terminals and explored the interpretation techniques of satellite remote sensing and UAV low-altitude mapping. This has led to the creation of a measurement system capable of real-time detection of changes in tunnel laying environments, significantly enhancing the ability to manage the impacts of large-scale subsidence in soft soil layers and construction activities along subway lines. Research and development outcomes, along with the equipment, have been implemented in several coastal cities, including Shanghai, Nanjing, Tianjin, Dongguan, and Wuxi.
(2) We have established a multi-source sensing integration platform to enhance the level of intelligent perception and early warning. Reconstructing the long-link and data synchronization technologies from front-end sensors to back-end servers, this research delves into data processing algorithms including outlier elimination, anomaly detection, and combined adjustments. It achieves remote control instruction encapsulation and integrated management for a wide array of multi-source devices such as laser rangefinders, inclinometers, survey robots, hydrostatic levels, electronic levels, crack meters, and fiber Bragg grating sensors. This approach addresses the application bottlenecks associated with the diversity of traditional monitoring sensors, the complexity of software, and data isolation, thereby meeting the all-weather remote monitoring requirements of critical subway tunnel sections. Consequently, it enhances the levels of intelligent perception, comprehensive data management, and early warning. The complete set of monitoring technology has the characteristics of high accuracy (sub millimeter), fast response speed (second), and diverse information, solving the needs of "large-scale, large-range, high-precision, automation" for subway tunnel monitoring. The results have achieved large-scale application of over 10,000 online equipment in subway structure monitoring and deep foundation pit monitoring sites in nearly 20 cities across China, such as Shanghai and Hangzhou.
(3) We have established a series of mobile rapid comprehensive detection equipment that can perceive the safety status of subway tunnel structures, achieving synchronous collection and real-time warning of multiple tunnel diseases. We have developed a lightweight mobile laser scanning robot with a total weight of 35 g, the fastest movement detection speed of m/h, and the highest image resolution of mm. Based on random consistency algorithm, we achieve real-time calculation and warning in high-speed acquisition (1 million points/second) scenarios; developed a mobile industrial camera disease collection equipment with professional lighting distribution design, and achieved crack monitoring accuracy of 0.1 mm; we have developed an adaptive algorithm for coupling

multiple diseases, achieving high-precision solution (1 mm level) for geometric deformations such as convergence and misalignment, and automatic extraction of multiple diseases such as structural leakage and concrete damage. The complete set of technology has broken through the technical bottleneck of mobile detection, achieving tunnel inspection from sampling style to loop by loop "CT scanning" physical examination, and improving operational efficiency by 200+ times, solving the "full coverage, high efficiency, and low cost" requirements of subway tunnel disease monitoring. The results have been promoted and applied in the detection work of over a thousand kilometers of subway tunnels in the entire subway network of Shanghai and Nanjing, as well as in more than 10 cities across the country.

References

Guo C (2014) Accuracy analysis of non-prism distance measurement of total station applied to tunnel convergence measurement. Urban Roads Bridges Flood Control 07:429–431

Wang R, Guo C, Chu W et al (2020) Monitoring and detection technologies for underground engineering operation and maintenance. Tongji University Press, Shanghai

Chapter 4
Ultimate Bearing Capacity of Tunnel Lining with Reinforcement

4.1 Introduction

With the rapid increase in the use of shield tunneling for subway projects, various structural issues related to shield tunnels have also emerged significantly in recent years. These issues include tunnel leakage, misalignment, cracking, crush or damage of tunnel segments, and overall tunnel deformations. Some of these issues have already affected the normal functioning of shield tunnels, leading to secondary problems that hinder normal operations, and in severe cases, compromise the structural safety and reliability of the tunnels. The primary reasons for these problems often lie in the insufficient understanding of the load-bearing properties of shield tunnel structures by those involved in construction. The limited experience and sometimes overly aggressive construction practices lead to inherent weaknesses in the tunnel's construction. In some cases, construction or engineering activities near existing subway lines exacerbate existing tunnel issues or introduce new structural challenges. Additionally, a lack of high-quality maintenance and repair efforts further accelerate the deterioration of these tunnels.

When issues arise in tunnels, such as tunnel leakage, deformation, and cracking or damage, it's critical to identify the point at which the tunnel structure becomes unstable or even dangerous. How can we assess the safety status of the structure? What constitutes the tunnel's maximum load-bearing capacity, specifically the maximum overburden and deformation the tunnel lining can tolerate? Determining the optimal timing for reinforcement and remediation is equally crucial. Which reinforcement and remediation methods should be employed, and what techniques and materials are best suited for these procedures? In the event of emergencies, what are the best immediate response measures?

To understand the safety status of shield tunnels, it is necessary to delve into the load-bearing mechanism of the tunnel lining structure, grasp comprehensively the

R. Wang et al., *Structural Safety Early Warning and Treatment Technology of Metro Tunnel in Soft Soils*, Key Technologies for Tunnel Construction under Complex Geological and Environmental Conditions,
https://doi.org/10.1007/978-981-96-6410-8_4

evolution of structural performance over time, investigate the influence of loading and unloading conditions caused by surrounding engineering activities, and propose maintenance and reinforcement methods. For this purpose, taking the common 15-m-deep segmentally assembled shield tunnel lining structure in Shanghai as an example, the research was conducted from 2011 to 2020, spanning nearly a decade. Utilizing Tongji University's structural testing platform, a total of 13 full-scale ring segment lining structure tests were carried out, systematically studying the mechanical evolution performance of different components and lining rings under various conditions. This includes exploring the structural morphology under different lateral pressure coefficients, the service status of longitudinal joint connection bolts, and the effects of different reinforcement materials and methods underground overload and peripheral unloading conditions.

From initial design to the fabrication of the loading devices and subsequent testing, we've developed a shield tunnel lining testing platform with comprehensive intellectual property rights. This platform allows for a holistic examination of the shield tunnel lining structures throughout their lifecycle. By analyzing the collected data meticulously, we discerned the load-bearing intricacies of shield tunnel linings. The insights garnered influenced structural designs and revealed the significance of the durability of longitudinal joint connecting bolts on overall structural performance. Building on this knowledge, we introduced a performance-based structural evaluation approach, segmenting the Shanghai shield tunnel structure into five distinct performance stages. This approach provides tailored maintenance recommendations for each stage. For the reinforcement phase, we proposed various reinforcement schemes and assessed their advantages and limitations through empirical testing. These research outcomes now serve as foundational references for the meticulous design and safe operational maintenance of shield tunnel lining structures.

4.2 Test Method

4.2.1 Test Specimen

As there would be thousands of possible operating conditions in the process of tunnel construction, it's impractical to test every potential situation. Based on the layout conditions of Shanghai's subway tunnels, commonly observed issues, and a series of considerations, we have focused on the effects of bolts, surcharge, lateral unloading, and force exposure after steel ring reinforcement on tunnel structural safety. We considered various scenarios: the presence, absence, or integrity of bolts, the magnitude of the lateral pressure coefficient corresponding to different scenarios, the size of the surcharge (simulating soil stacking), and the unloading on one or both sides of the tunnel (simulating pit excavation, tunnel leakage, etc.). Furthermore, considering the combination of reinforcement materials and the scope of reinforcement, we finally selected 13 shield tunnel lining rings for typical scenario testing.

These specimens are modeled after the lining rings of Shanghai subway tunnels. Due to different testing objectives, variations exist in bolt conditions, reinforcement materials, reinforcement ranges, track bed, and strutting arrangements, as shown in Tables 4.1 and 4.2. The prototype lining ring dimensions are an external diameter of 6.2 m, an internal diameter of 5.5 m, a segment thickness of 0.35 m, and a ring width of 1.21 m. Each ring is divided into six sections, including one top seal segment (F segment), two adjacent segments (L1 and L2), two standard segments (B1 and B2), and one bottom seal segment (D segment). The segments are connected using M30 straight bolts, as illustrated in Fig. 4.1.

Table 4.1 Test specimens of the unreinforced lining ring

Number	Bolt condition	Lateral pressure coefficient	Simulated condition	Diagram
1	Normal	0.50	Ground overload	Figure 4.1
2	Normal	0.65	Ground overload	Figure 4.1
3	Normal	0.70	Peripheral unloading	Figure 4.1
4	No bolts	0.65	Ground overload	–
5	Bolt corrosion rate 10%	0.65	Ground overload	Figure 4.1
6	No bolts	0.70	Peripheral unloading	–
7	Bolt corrosion rate 10%	0.70	Peripheral unloading	Figure 4.1

Table 4.2 Test specimens of the reinforced lining ring

Number	Reinforcement material	Reinforcement range	Simulated condition	Diagram
1	Steel plate	Full ring, 85 cm width	Ground overload	Figure 4.2
2	Steel plate	Full ring, top 66° area 110 cm width, other 85 cm	Peripheral unloading	Figure 4.3
3	Steel plate	282° large half-ring (excluding track bed range), 85 cm width	Ground overload	Figure 4.4
4	Steel plate	218° large half-ring (reduced bottom range), 85 cm width	Ground overload	Figure 4.5
5	Composite cavity (standard arrangement)	6 pieces 100° + 4 pieces 95° + 4 pieces 95° (cavity through-joint arrangement)	Ground overload	Figure 4.6
6	Composite cavity (reinforced arrangement)	4 pieces 100° (2 pieces 126°) + 4 pieces 100° + 4 pieces 100° (cavity partial overlap)	Peripheral unloading	Figure 4.7

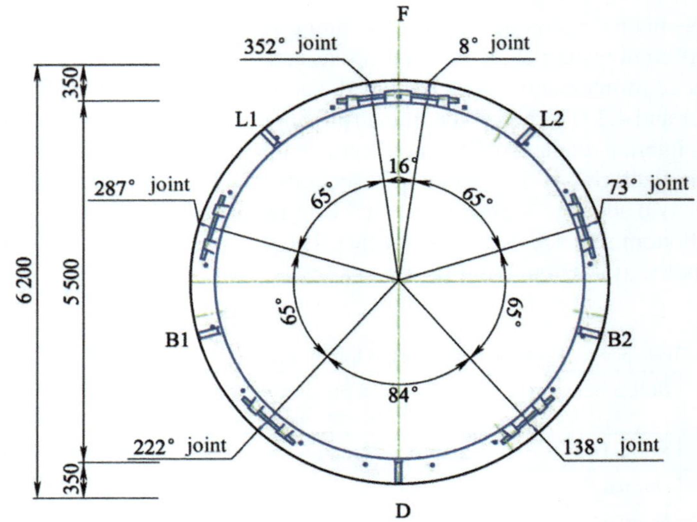

Fig. 4.1 Schematic of the Shanghai Metro tunnel lining structure ring (unit: mm)

The reinforcement test simulates the process of secondary loading of the real tunnel structure. At the initial stage of the loading test, the specimen represents the lining ring of the Shanghai subway tunnel (Fig. 4.1). When the structural convergence displacement reaches 2% D (D is the tunnel's external diameter), the tunnel lining structure is reinforced while maintaining the applied load, forming the reinforced lining ring specimen, as shown in Figs. 4.2, 4.3, 4.4, 4.5, 4.6 and 4.7.

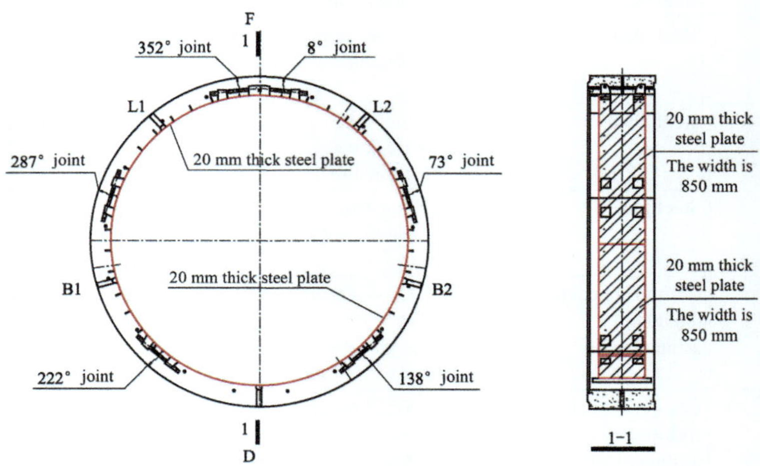

Fig. 4.2 Schematic of full-circle steel plate reinforcement (85 cm wide)

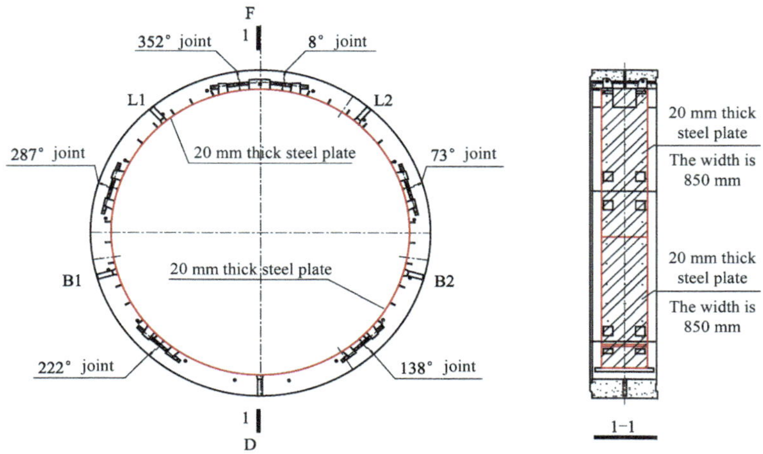

Fig. 4.3 Schematic of full-ring steel plate reinforcement (top 66° area 110 cm wide, remainder 85 cm wide)

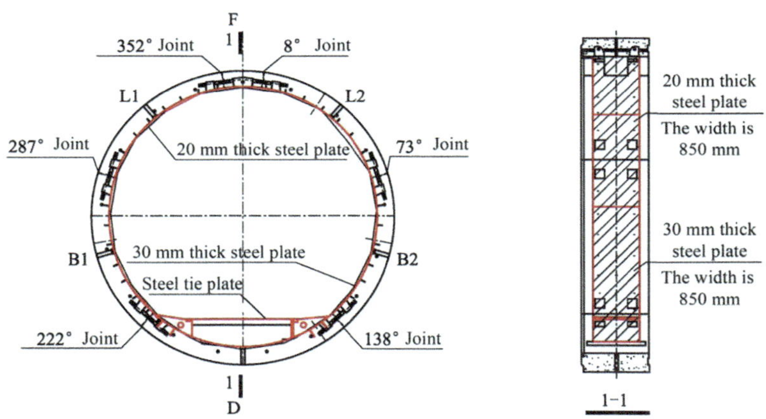

Fig. 4.4 Schematic of 282° steel plate reinforcement

4.2.2 Loading System

The experiment is conducted with 24 jacks installed at the loading points (as illustrated in Fig. 4.8). The jack forces are transferred to the central steel ring through the loading beams, establishing a self-balancing loading system. Each loading point comprises a load distribution beam, a load-bearing beam, and two steel tension rods, as depicted in Fig. 4.9. The reaction device can deliver a maximum horizontal load of 100 T per point and allows for a maximum displacement of 400 mm per point. While horizontal loading offers convenience in experimental procedures, the friction

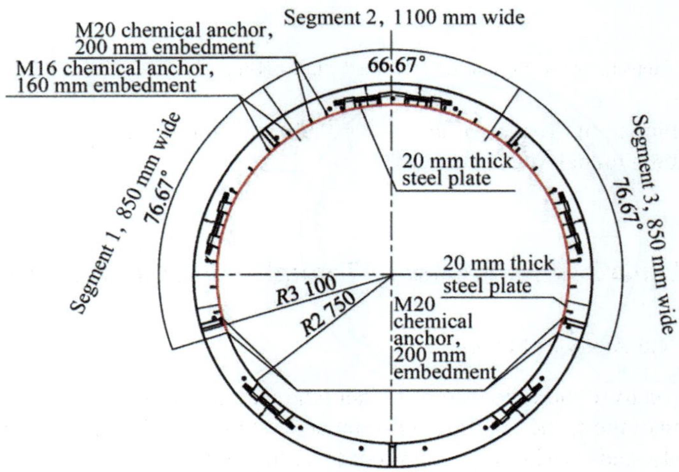

Fig. 4.5 Schematic of 220° steel plate reinforcement

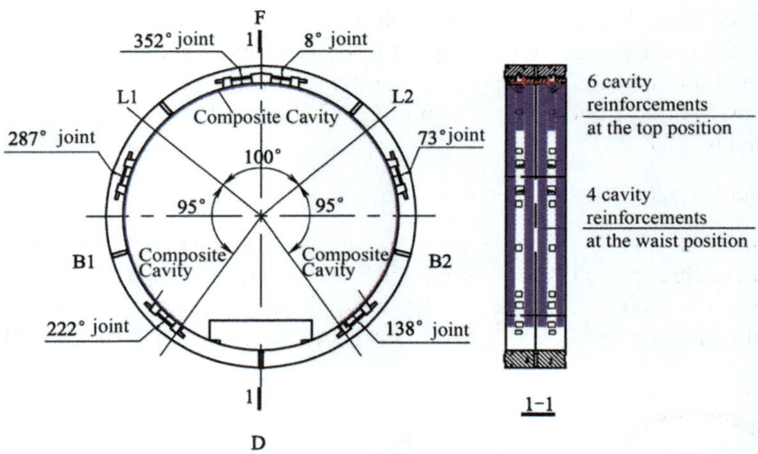

Fig. 4.6 Schematic of composite cavity (standard) reinforcement

between the specimen and the steel base plate can influence the planar deformation state. To mitigate this friction, a steel base plate is uniformly positioned beneath the entire specimen. To ensure minimal frictional resistance, 150 steel balls are strategically placed between the bottom of the specimen and the steel base plate, creating a low-resistance rolling support condition.

In the reinforcement test, the unreinforced tunnel segment is first loaded to a reinforcement state with the top–bottom displacement of 12 cm. Subsequently, the deformation is locked, and the segment is reinforced. Once the reinforcement process

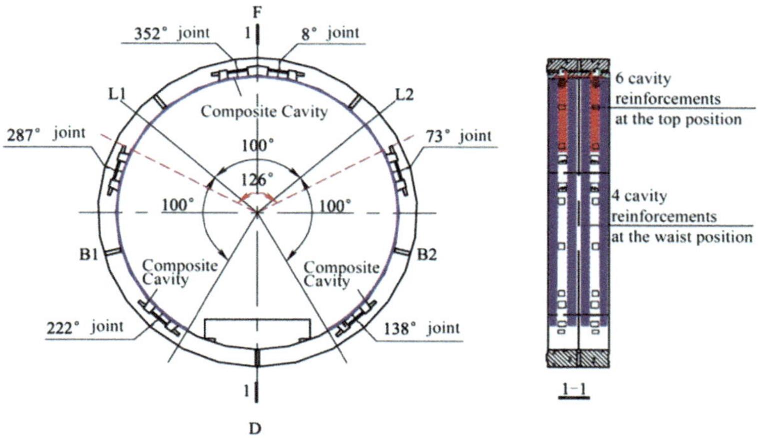

Fig. 4.7 Schematic of composite cavity (reinforced) reinforcement

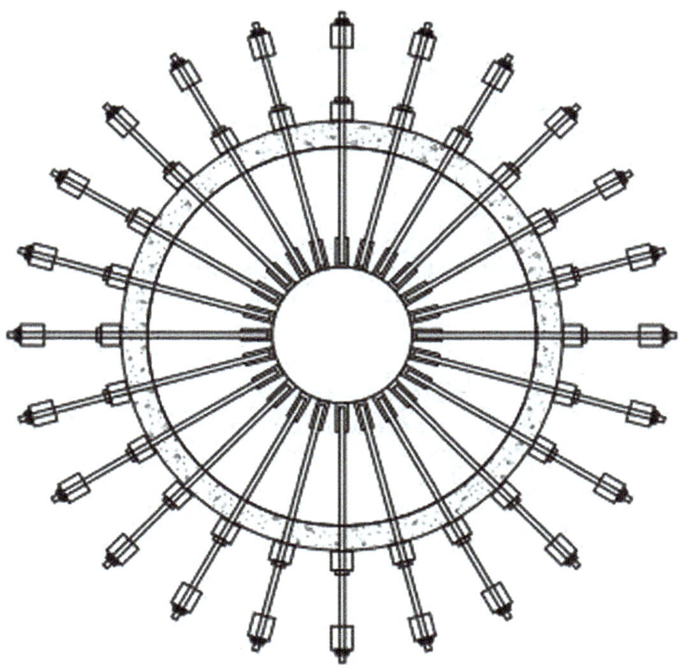

Fig. 4.8 Schematic layout of the loading apparatus

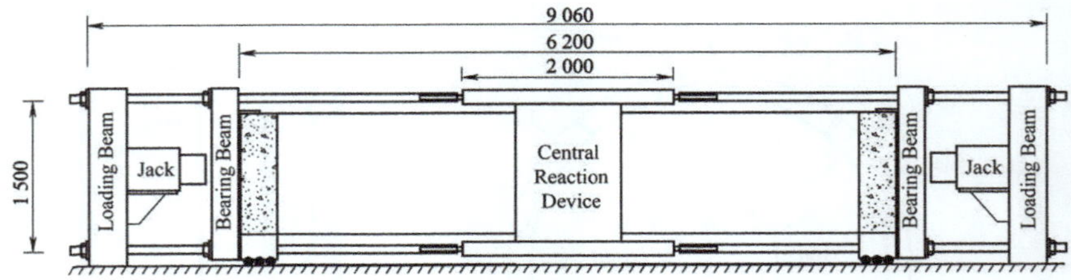

Fig. 4.9 Schematic cross-sectional view of the loading apparatus (units: mm)

Fig. 4.10 Deformation retaining device

is complete, the segment undergoes a second round of loading. This procedure simulates the practical reinforcement of a deformed tunnel. For this process, a deformation holding device is utilized, as illustrated in Fig. 4.10. Upon reaching the designated reinforcement state during testing, the deformation is fixed in place using bolts. However, during the actual reinforcement phase, these bolts are released to allow for continuous loading. This device serves a dual purpose: (1) It acts as a cushioning beam during loading, ensuring a consistent vertical load distribution, and (2) when not in active use, it functions as a passive reaction beam, preserving the test specimen's deformation state.

4.2.3 Load Design

In the experiment, the loads were divided into three groups. Within each group, the value of the load at each point was the same, and the loading was fully synchronized. As shown in Fig. 4.11, the load grouping is as follows: Group P_1 has 6 loading points; Group P_2 has 10 loading points; and Group P_3 had 8 loading points.

In the full-scale test for the overload scenario, the inter-group load relationship is maintained as $P_2 = 0.65 \times P_1$ and $P_3 = (P_1 + P_2)/2$. When the lateral earth pressure load P_2 reaches its maximum value of 275 kN, it remains constant, while the external

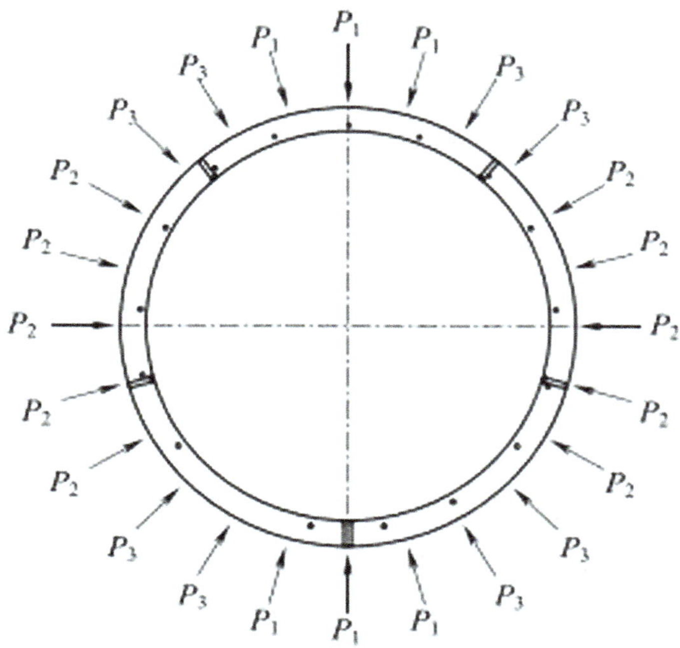

Fig. 4.11 Circumferential load grouping

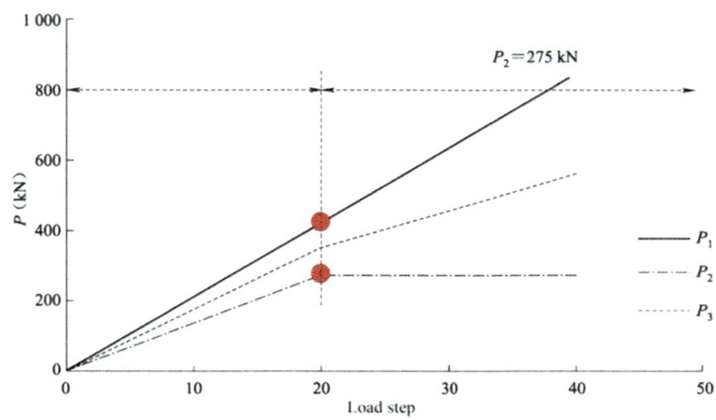

Fig. 4.12 Load curve under overload conditions

load P_1 continues to increase. Despite this increase, the relationship $P_3 = (P_1 + P_2)/2$ is still satisfied until the tunnel lining segment fails. The load curve is shown in Fig. 4.12.

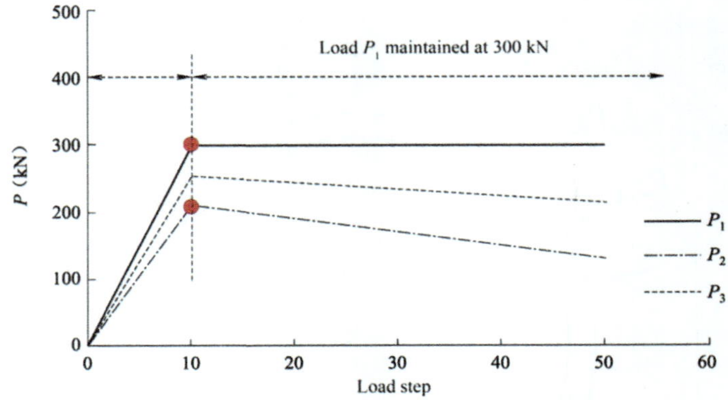

Fig. 4.13 Load application curve for unloading conditions

During the full-scale unloading test, the load relationship between the groups satisfies $P_2 = 0.7 \times P_1$ and $P_3 = (P_1 + P_2)/2$. Considering that the tunnel segment is buried at a depth of 15 m, based on the aforementioned calculated standard value of external load under tunnel working conditions, once the top and bottom load P_1 reaches 300 kN, it remains constant. Meanwhile, the lateral earth pressure starts to decrease continuously. P_3 still satisfies the relation $P_3 = (P_1 + P_2)/2$ until the tunnel segment is damaged. The load-loading curve is shown in Fig. 4.13.

4.2.4 Test Content and Methods

During the experimental process, the primary metrics measurements include overall deformation, joint opening, circumferential bolt strain, and segment strain. For the reinforcement tests, there's an additional measurement of strain in the reinforcement and the bond slip between the reinforcement and the segment. Observations during the tests capture the progression of surface cracks and joint damages on the segment under various loading conditions. The arrangement of monitoring points is based on preliminary results from the finite element analysis, and are strategically positioned at locations indicating significant computational values. Additional points are systematically positioned in other areas. The selection of instrumentation is also in line with the preliminary results of the finite element analysis, ensuring that the chosen tools meet the requisite range and precision demands.

A significant aspect of the tests is monitoring the deformation of the segment structure. This is gauged using 14 wire displacement sensors placed on the segment's cross-sectional profile, as depicted in Fig. 4.14. Excluding sensors 2 and 14, the remaining 12 sensors are paired to form six sets of opposite measurement points. This configuration enables the capture of the segment's radial deformation from

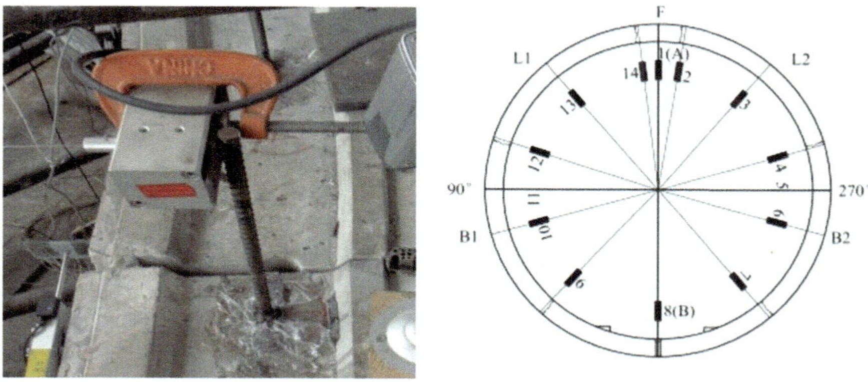

Fig. 4.14 Schematic layout of measurement points for overall deformation

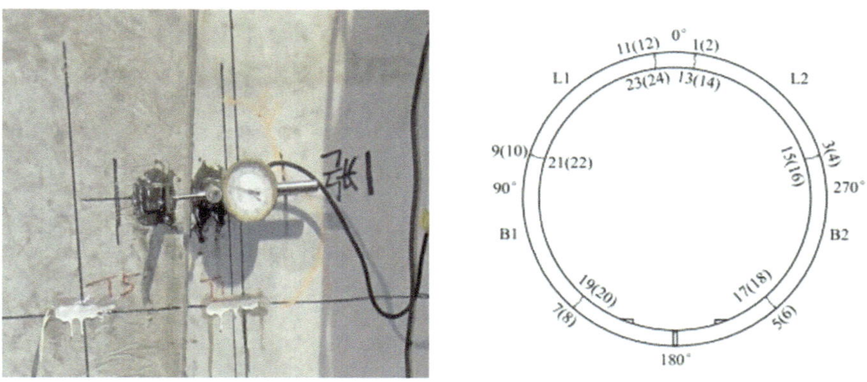

Fig. 4.15 Schematic layout of measurement points for joint opening

various angles, offering a comprehensive insight into the segment's deformation characteristics.

The opening of the joints is measured by displacement sensors, with 2 measurement points each arranged on the inner and outer surfaces, as shown in Fig. 4.15.

The bolt strain is measured using pin-type strain gauges, with the model BX3AA-120. Prior to affixing the strain gauges, it's essential to carve grooves into the bolts. The strain gauges are then adhered within these grooves and sealed using adhesive. In total, there are six circumferential joints. Each joint incorporates two Grade 5.8 M30 circumferential bolts, culminating in 12 bolts overall. For each bolt, two symmetrical strain measurement points are established to monitor the variations in the bolt's load during the test, amounting to a collective 24 measurement points, as illustrated in Fig. 4.16.

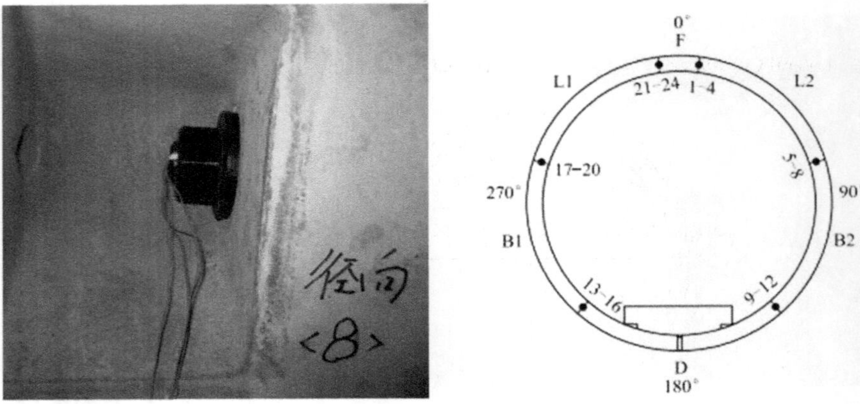

Fig. 4.16 Schematic layout of the strain measurement points on the circumferential bolts

The strain on the tunnel segment's surface is measured using the DX50AA-120 strain gauge with a gauge length of 50 mm. For the strain measurement on the surface of the reinforcement material, the BX120-3AA strain gauge is employed, with a gauge length of 3 mm. Strain measurement points are uniformly arranged around the circumference, with an increased number of points positioned near critical sections, as shown in Fig. 4.17.

In the experiment, an electronic displacement sensor with a precision of 0.01 mm is used to measure the adhesive slip between the tunnel lining and the reinforcement body. To set up the measurement, the base of the displacement sensor is first anchored to the tunnel lining using epoxy resin. In a circumferential direction, a glass barrier is secured onto the composite chamber. The probe of the displacement sensor is then placed against the glass barrier. When relative slippage occurs between the tunnel lining and the composite chamber, the displacement sensor records the relative tangential slip distance between them. The layout of the adhesive slip measurement points is illustrated in Fig. 4.18.

Besides crack observations, whether it's bolt strain, surface strain of the segment and reinforcement material, or joint opening, the static strain electric measurement method is employed. The mechanical quantities (such as strain, displacement, etc.) measured from the specimen are converted into corresponding electrical signals by the rosette strain gauges and electronic displacement sensors connected to the DH3816 static strain data collector.

The complete data collection system comprises sensors, the DH3816 static data collector, a computer, and supporting software. The measurement range of the testing instrument for strain is from -2×10^4 to 2×10^4 $\mu\varepsilon$. The resolution is 1 $\mu\varepsilon$, zero drift is no more than 4 $\mu\varepsilon$/h, and system uncertainty is less than 0.5% or $\pm$ 3 $\mu\varepsilon$. The entire testing instrument is characterized by accuracy, reliability, and speed (with 400 measurement points collected within 10 s).

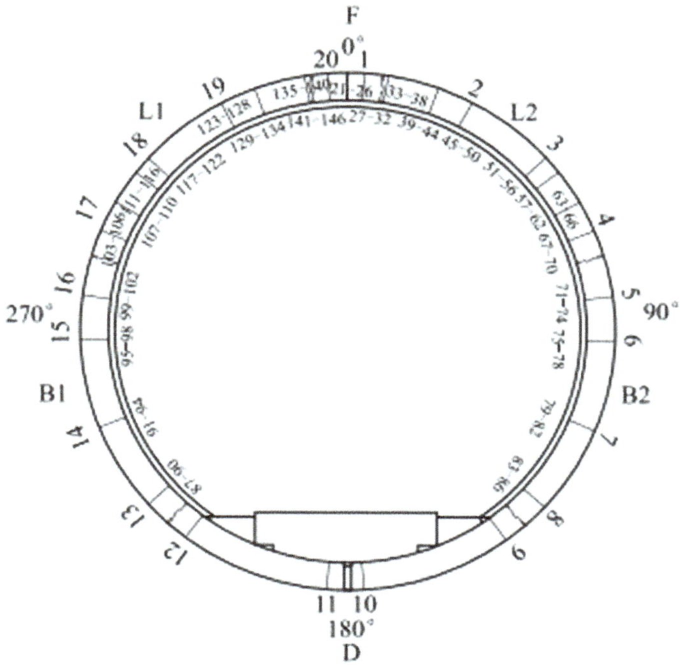

Fig. 4.17 Schematic layout of the strain measurement points on the tunnel lining and reinforcement surface

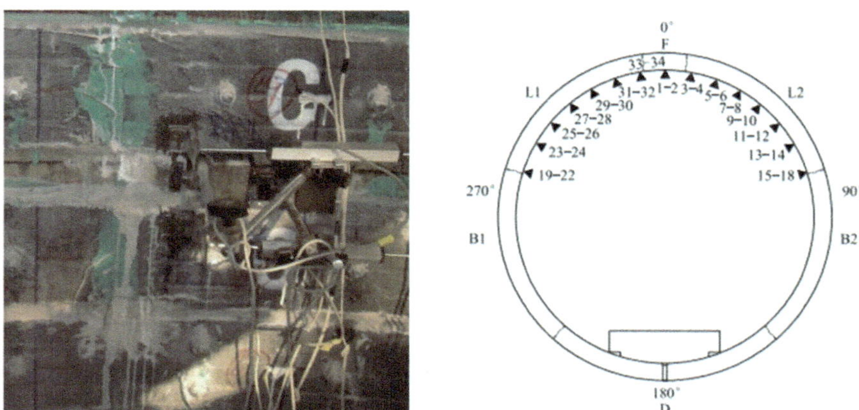

Fig. 4.18 Schematic layout of the measurement points for adhesive slippage between the segment and composite cavity

4.3 Test Results and Analysis

From Tables 4.1 and 4.2, it's evident that a range of full-scale experimental studies on shield tunnel lining structures considered multiple factors: three lateral pressure coefficients (0.5, 0.65, and 0.7), three bolt conditions (normal, rusted, and failed), two main types of test specimens (unreinforced and reinforced), two reinforcement materials (steel plate and composite cavity), four reinforcement methods (full ring reinforcement, track bed reinforcement with 28 and 220° reinforcements, and top strengthening), and two loading methods (top overload and peripheral unloading). When all these factors were combined, the resulting working conditions numbered in the hundreds. The full-scale indoor tests chose scenarios most likely to occur in actual operations. By selectively contrasting the typical conditions simulated in these tests, the primary conclusions drawn addressed the six aspects: lateral pressure coefficient, bolt condition, reinforcement presence, reinforcement method, reinforcement material, and the method of loading/unloading.

4.3.1 Influence of the Lateral Pressure Coefficient

The primary difference between the tests of Ring 1 and Ring 2 is the variation in lateral earth pressure coefficients simulated during the loading process by the jack load. For Ring 1, the coefficient was 0.5, while for Ring 2, it was 0.65. The representative load–displacement curves derived from the full-scale tests for both rings are presented in Figs. 4.19 and 4.20. Both tests exhibited similar failure patterns, with damage predominantly localized at the joints. The first sign of localized damage manifested in the concrete of the core area at the top joint followed by compressive fracturing of the concrete at the outer edge of the top joint. Ultimately, the concrete at the inner edge of the springline joint underwent compressive failure. Damage at four joint locations led to the formation of a plastic hinge, preventing the structure from bearing additional load. The final states of the tests for both rings are showcased in Figs. 4.21 and 4.22.

The sequence and cause of the appearance of critical performance points in the two structures are consistent. The effect of the lateral pressure coefficient is reflected in the differing timings of when these critical performance points arise. A larger lateral pressure coefficient indicates stronger structural load-bearing capacity. This is because the lateral pressure coefficient represents the degree of uniformity of the radial pressure applied to the circular structure. A larger coefficient implies a more uniform radial pressure, allowing the circular structure to predominantly bear axial pressure in response to this uniform radial load. This maximizes the material properties of the reinforced concrete tunnel lining.

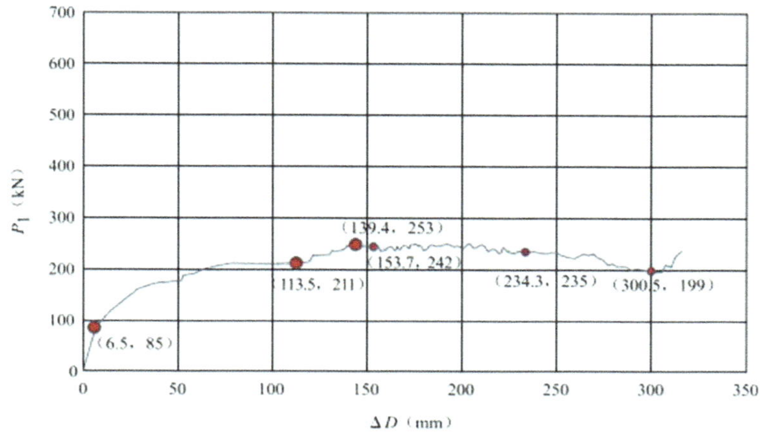

Fig. 4.19 Load–displacement curve for the first ring ($K = 0.5$)

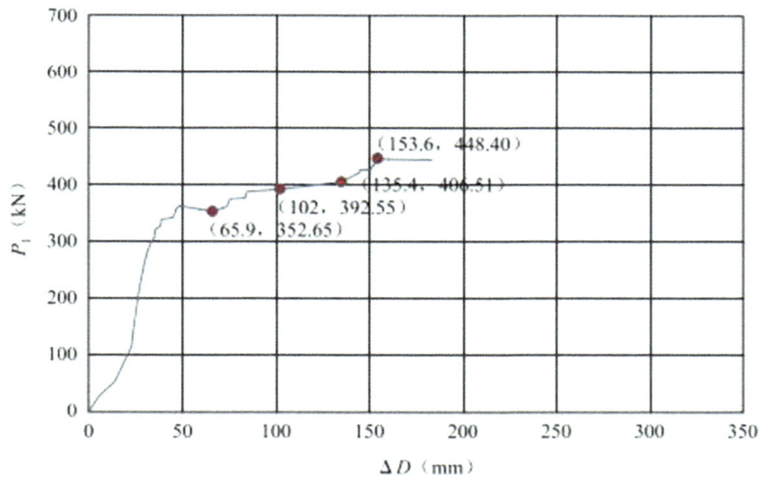

Fig. 4.20 Load–displacement curve for the second ring ($K = 0.65$)

4.3.2 Impact of Circumferential Bolts

Circumferential bolts are often used as positioning devices in Europe, temporarily fixing the tunnel lining segments. Once the segments are assembled into a ring, they are removed. However, in Asian regions (such as China and Japan), circumferential bolts are considered essential permanent components in the circumferential joint of tunnel linings. Thus, there is no unified global understanding about the role of circumferential bolts in the shield tunnel lining structure. Moreover, bolt corrosion

Fig. 4.21 Status after the end of the test for the first ring ($K = 0.5$)

Fig. 4.22 Status after the completion of the test for the second ring ($K = 0.65$)

in operational tunnels is a common issue, and how to evaluate the impact of bolt corrosion on the lining structure has always been a longstanding scientific challenge in the safety and operational maintenance of shield tunnels.

To clarify the role of circumferential bolts, six full-scale tests were conducted, namely the 2nd to 7th rings. According to the differences in load conditions, they were divided into two groups: The overload conditions included the 2nd (with non-corroded bolts), 4th (without bolts), and 5th (with corroded bolts) rings, as shown in Fig. 4.23. The unloading conditions included the 3rd (with non-corroded bolts), 6th (without bolts), and 7th (with corroded bolts) rings, as shown in Fig. 4.24. The corroded bolts mentioned here refer to bolts with a 10% corrosion rate, as depicted in Fig. 4.25. After comparative analysis of the test results, the following main conclusions were drawn.

(1) Under the conditions of top overload and peripheral unloading, the structural response of tunnel lining sections with both intact and corroded circumferential bolts consistently displays three distinct phases: the elastic phase, the elastoplastic phase, and the plastic phase. In contrast, sections with failed circumferential bolts lack a pronounced elastoplastic phase, reaching their limit state rapidly after the onset of initial local damage.
(2) The role of circumferential bolts is primarily to influence the bending stiffness of the joints, which, in turn, affects the overall stiffness of the structure. The degree of this influence is directly correlated with the structural loading process and is evident in distinct stages. Elastic Phase: Even though the bolts

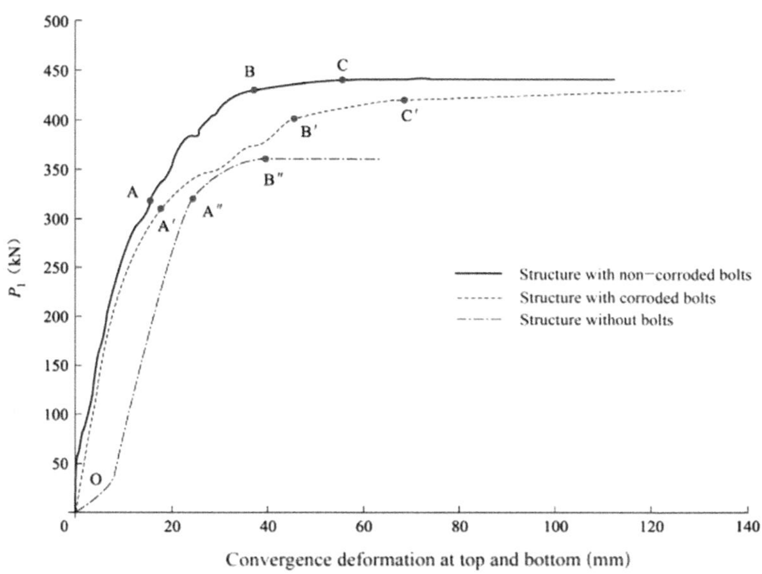

Fig. 4.23 Load–displacement curves under overload conditions for the 2nd ring (with non-corroded bolts), 4th ring (without bolts), and 5th ring (with corroded bolts)

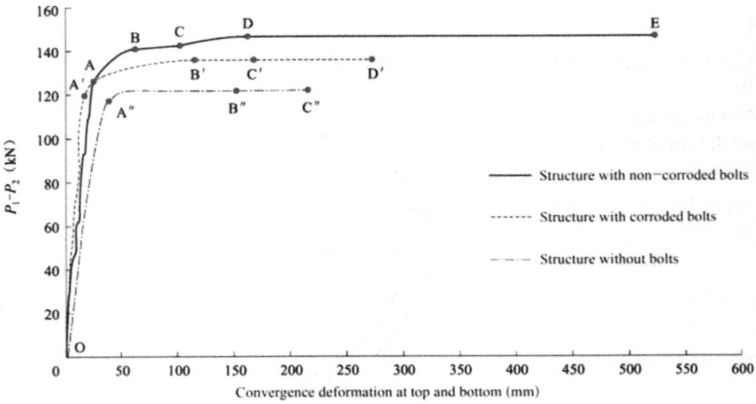

Fig. 4.24 Load–displacement curves under unloading conditions for the 3rd ring (with non-corroded bolts), 6th ring (without bolts), and 7th ring (with corroded bolts)

are subjected to stress, the force exerted is relatively minor. Corrosion of the bolts has minimal effect on the stiffness of the joint and overall structure. However, bolt failure results in a noticeable reduction in both joint and structural stiffness. Elastoplastic Phase: During this phase, the concrete in the top joint's core area undergoes compression damage. Bolts in each joint play a significant role. Their tensile and compressive stiffness during this phase directly affects the bending stiffness of the joint, and consequently, the overall structural stiffness. Plastic Phase: The ultimate state of the joint is marked by edge concrete crushing. The corrosion and subsequent failure of the circumferential bolts can lead to a premature onset of this crushing, diminishing the structure's maximum load-bearing capability.

(3) Damage Mechanism in Structures with Corroded Circumferential Bolts: Damage is predominantly localized at the joints. The initial localized damage invariably results from the compression failure of the concrete in the core areas of the 8 and 352° joints. This is followed by the crushing of the outer-edge concrete of these joints. The final stage sees the crushing of the inner-edge concrete of the 73 and 287° joints. These sequential damages cause the tunnel lining structure to adopt a geometrically variable mechanism, leading it to its load-bearing limit. Due to the diminished tensile and compressive stiffness of the corroded bolts, the joints experience increased deformation when the bolts yield. In the overload scenario, no observable bolt yielding was noted, whereas, under unloading conditions, only a few bolts exhibited yielding.

(4) Damage Mechanism in Structures with Failed Circumferential Bolts: As soon as there's damage in the top joint's core area, each joint effectively becomes a structure with zero bending stiffness. This transformation results in the entire tunnel lining structure evolving into a geometrically adaptable mechanism, exhibiting a

Fig. 4.25 Corroded bolts used in the full-circle lining structure

sharp escalation in overall deformation and becoming incapable of further load-bearing. In essence, structures equipped with circumferential bolts undergo a failure due to strength, while those with failed bolts experience a failure due to loss of stiffness.

4.3.3 Reinforcement Decision

On the basis of a full understanding of the bearing mechanism and failure mechanism of the shield tunnel lining structure, a safety evaluation method for shield

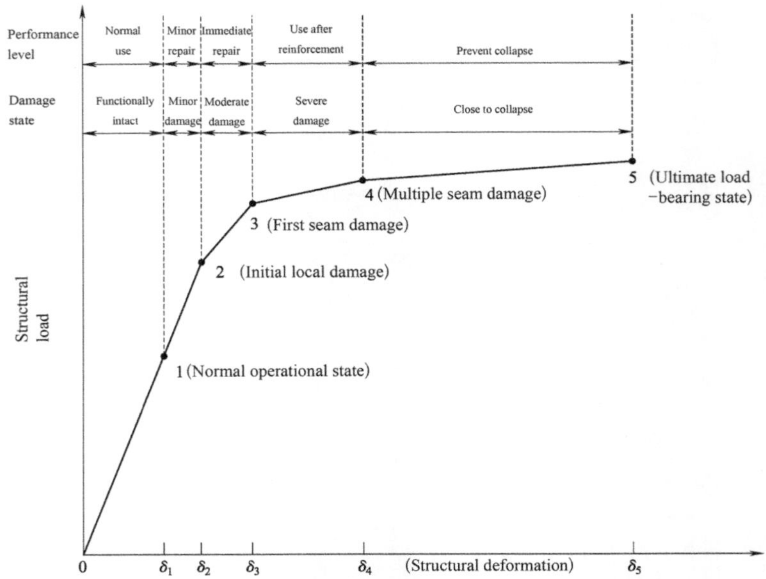

Fig. 4.26 Division of performance levels for shield tunnel lining structure

tunnels based on performance levels has been proposed (Liu et al. 2013). This method uses horizontal convergence displacement as a representative indicator, dividing the segmented shield tunnel structure in Shanghai into five performance level stages, as shown in Fig. 4.26.

Based on the onset of damage to the initial joint of the tunnel lining structure and prior to damage extending to multiple joints, it is advised to undertake reinforcement measures for the lining structure. To assess the influence of such reinforcement on the structural performance, a common engineering approach using steel plate reinforcement was employed. Reinforcement measures were initiated for the entire ring when the vertical convergence displacement of the structure reached $2\%D$, as detailed for the 8th ring in Table 4.2.

The representative load–displacement curve acquired from the test is depicted in Fig. 4.27. Upon reaching a vertical convergence displacement of 127 mm, the lining structure was reinforced with internal steel plates. Subsequent to this reinforcement, although the external load on the structure increased significantly, the rate of displacement growth decelerated. This observation underscores the efficacy of internal steel plate reinforcement in augmenting both the load-bearing capacity and stiffness of the structure. As the load escalated to 586 kN, a bonding failure between the steel plate and the original tunnel segment emerged at the apex and base of the structure. Following this, the structure rapidly lost its load-bearing capability, with the entire failure mode being brittle in nature. The post-test condition is illustrated in Fig. 4.28. When contrasted with the second ring that wasn't reinforced, there was an enhancement of 31% in the ultimate load-bearing capacity.

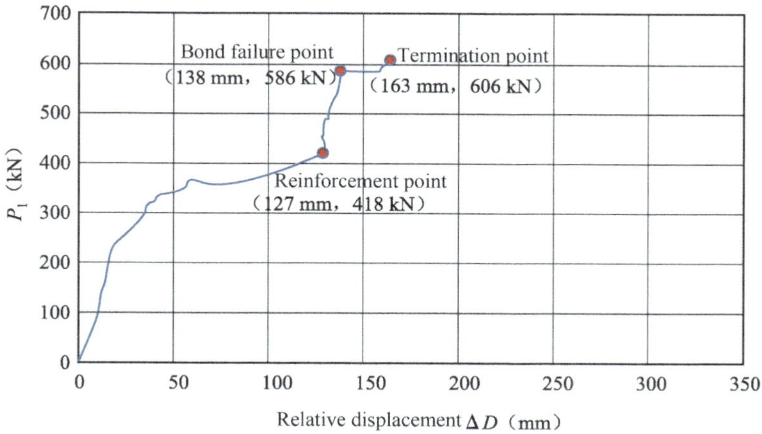

Fig. 4.27 Load–displacement curve for the 8th ring (whole ring steel plate reinforcement test)

Fig. 4.28 Image of the status after the test of Ring 8 (steel plate full ring reinforcement experiment)

4.3.4 Impact of Reinforcement Methods

The test results of the 8th ring have confirmed the effectiveness of the full-ring steel plate reinforcement. However, there are certain flaws in the failure mode and reinforcement technique, for which three optimization designs were conducted.

The first optimization approach took into consideration that the ultimate bearing capacity of the full-ring steel plate reinforcement structure is controlled by the bond failure between the top steel plate and the tunnel lining segment. One of the optimization ideas was to widen the top steel plate and strengthen the bond between the top steel plate and the original concrete lining, as shown in Fig. 4.3. Subsequently, the 9th ring test study was further conducted. The typical load–displacement curve is shown in Fig. 4.29, and the final state after the test is shown in Fig. 4.30. Compared to the 8th ring test, the difference lies in, on one hand, the widened width of the top steel plate in the 9th ring test, and on the other hand, the 9th ring adopted a peripheral unloading condition. Therefore, the two do not have direct quantitative comparison conditions. However, from a qualitative perspective, the failure modes of both are the bond failure between the top steel plate and the original tunnel lining segment. Therefore, the optimization idea of widening the top steel plate did not have an effect on changing the failure mode of the reinforced structure. In the later stages, one should consider optimization based on strengthening the bond strength of the interface.

The second optimization was designed considering that in actual operational tunnels, the bottom segments are often covered by a track bed. Given the need for convenient reinforcement during construction, two further modifications were

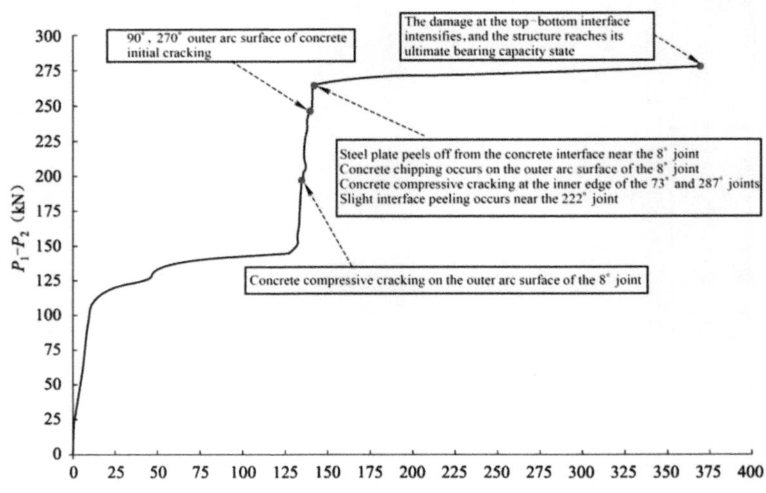

Fig. 4.29 Load–displacement curve of the 9th ring (whole ring steel plate with widened top)

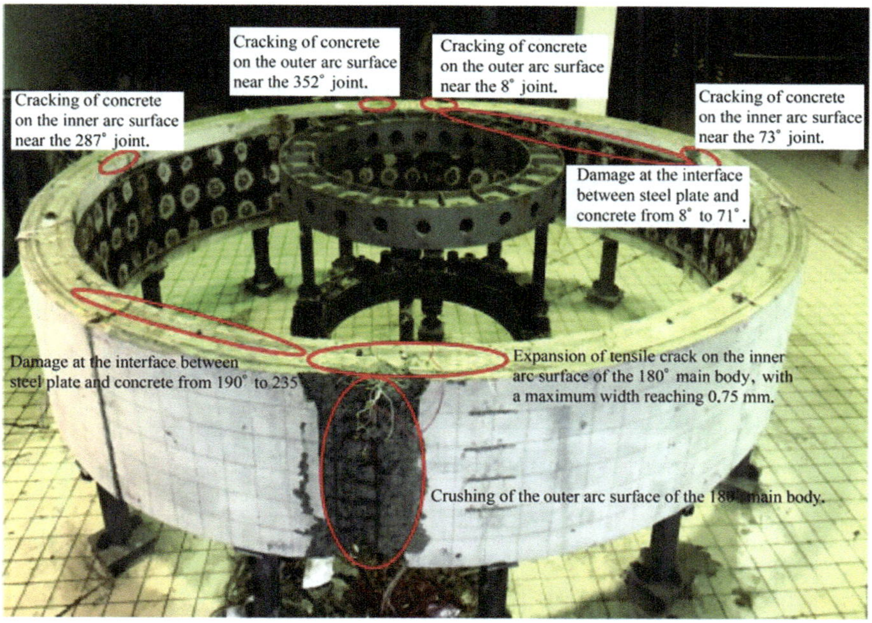

Fig. 4.30 Post-experiment condition of the 9th ring (widened steel plate at the top of the entire ring)

introduced. In the first modification, the track bed area was excluded from the steel bonding process. Instead, corbels were used to connect steel plates near the track bed, as illustrated in Fig. 4.4, representing the 10th ring. The standard load–displacement curve for this setup is represented as a dashed line in Fig. 4.31. The end-state after the test can be seen in Fig. 4.32. When compared with the whole ring reinforcement seen in the 8th ring (Fig. 4.27), both reinforcement methods result in the same structural failure mode. While the semi-ring reinforced structure shows marginally lesser overall structural integrity, the reduction in its ultimate load-bearing capacity is no more than 5%.

The third optimization stemmed from the realization that the construction of the corbels near the track bed was intricate and time-consuming, hindering the efficiency of internal tension steel ring reinforcement. Therefore, the design was further refined to eliminate the corbels, reducing the steel plate reinforcement to a 220° segment, as depicted in Fig. 4.5. This design corresponds to the 11th ring. The representative load–displacement curve for this design is the solid line in Fig. 4.31, and its end state after testing is presented in Fig. 4.33. Removing the corbel design notably improved construction efficiency. Though there was a decrease in the structure's rigidity, the ultimate load-bearing capacity remained essentially the same.

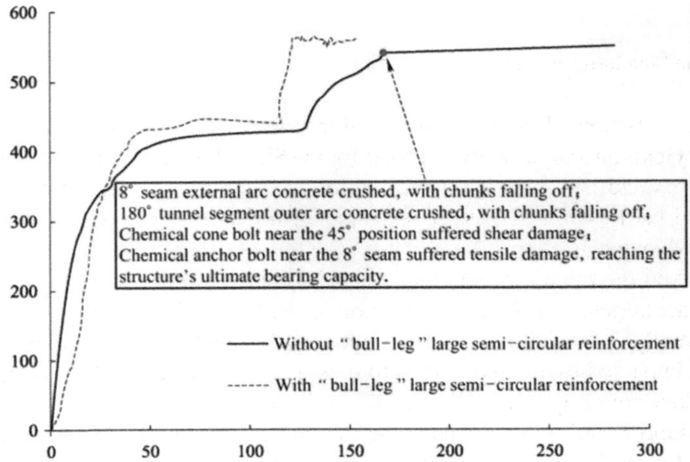

Fig. 4.31 Load–displacement curve for the 10th ring (large semi-circular steel plate + corbel structure) and the 11th ring (large semi-circular steel plate without corbel structure)

Fig. 4.32 State after the test of Ring 10 (large semi-circular steel plate + corbel)

4.3.5 *Influence of Reinforcement Materials*

While the internal tension steel ring reinforcement technique is relatively established, it grapples with challenges like the excessive weight of the steel plate and elevated

Fig. 4.33 State after the test of Ring 11 (large semi-circular steel plate without corbel)

reinforcement costs. These challenges hinder efficient and rapid reinforcement for operational subway tunnels. To address these challenges, an innovative approach in shield tunnel reinforcement has been proposed: the composite cavity reinforcement method.

This composite cavity component consists of steel, carbon fiber, and high-strength mortar, with a cross-sectional dimension of 160 mm × 40 mm. Initially, the cavity is pre-formed as a steel hollow section. Carbon fiber is then adhered to the steel surface through composite material fabrication techniques such as vacuum molding, heating, and hand lay-up. This process results in a lightweight and high-strength composite cavity. Once the composite cavity is affixed to the inner surface of the shield tunnel, it is subsequently filled with high-strength mortar. From a construction methodology standpoint, this process is both straightforward and efficient. Its reinforcing efficacy was further explored through comprehensive tests.

Two full-scale indoor tests were conducted for the composite cavity reinforcement, namely Ring 12 and Ring 13. The distinction between the two tests lies in the arrangement of the composite cavities, with designs referred to as basic reinforcement and enhanced reinforcement. The standard reinforcement approach is depicted in Fig. 4.6. Here, the composite cavity is divided into three sections for reinforcement. At the top position, six composite cavities are used for reinforcement, spanning a central angle of 100°. For the left and right mid-sections, four composite cavities are employed, each covering a central angle of 95°. The four composite cavities in the mid-section are interconnected at the top position using an internal splice joint, which is then sealed with a structural adhesive. The enhanced reinforcement design is illustrated in Fig. 4.7. In this method, the composite cavity is also divided into three sections. When compared to the standard reinforcement, there is an addition of two separate composite cavities at the top seal block position, extending the central angle from the original 100 to 126°. Furthermore, the length of the four composite cavities in the left and right mid-sections has been increased, expanding their central angle from the initial 95 to 100°.

The overall load–displacement curves comparing the two sets of composite cavity-reinforced tunnel structures are shown in Fig. 4.34. When comparing the steel plate (Fig. 4.27) and the composite cavity as reinforcement materials, their reinforcement effects are essentially equivalent. In terms of the damage pattern, the standard reinforcement method experienced damage at the top sealing block position of the composite cavity due to insufficient anchoring, causing it to pop out. The state of the tunnel after the test for this method is depicted in Fig. 4.35. Conversely, the enhanced reinforcement method increased the anchoring length at this position, preventing any damage during the test. Its post-test condition is shown in Fig. 4.36. Considering the initial point of damage, the standard reinforcement method experienced adhesive failure within the range of 350–100°. On the other hand, the enhanced reinforcement method improved the adhesive performance between the tunnel segment and the composite cavity. Its initial damage was due to the concrete crushing at the edge of the 35° joint rather than local adhesive failure. In terms of load-bearing capacity, both reinforcement methods significantly enhance the tunnel structure's ultimate load-bearing capacity and overall stiffness. The standard reinforcement and enhanced reinforcement methods improved the prototype structure's ultimate load-bearing capacity by 23.1% and 36.8% respectively. Meanwhile, the increase in overall stiffness was 1621.4% for the standard method and 2070.0% for the enhanced method. The enhanced reinforcement method outperforms the standard method, proving its optimization to be effective. This provides a valuable reference and experimental basis for actual engineering practices of composite cavity reinforcement.

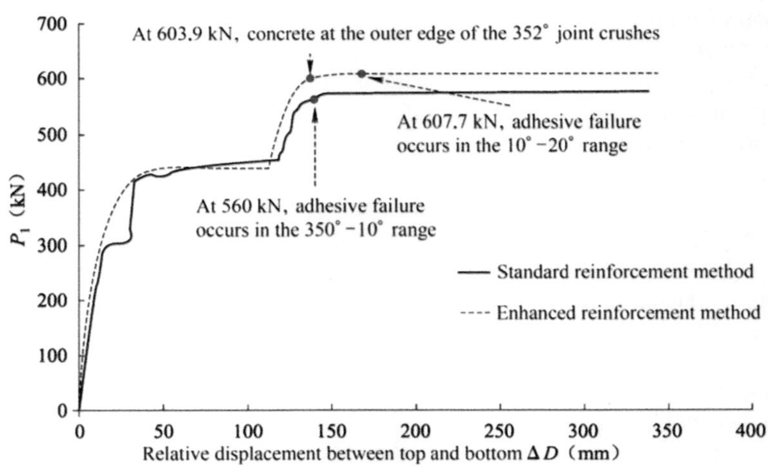

Fig. 4.34 Load–displacement curve of the 12th ring (standard arrangement) and the 13th ring (enhanced arrangement)

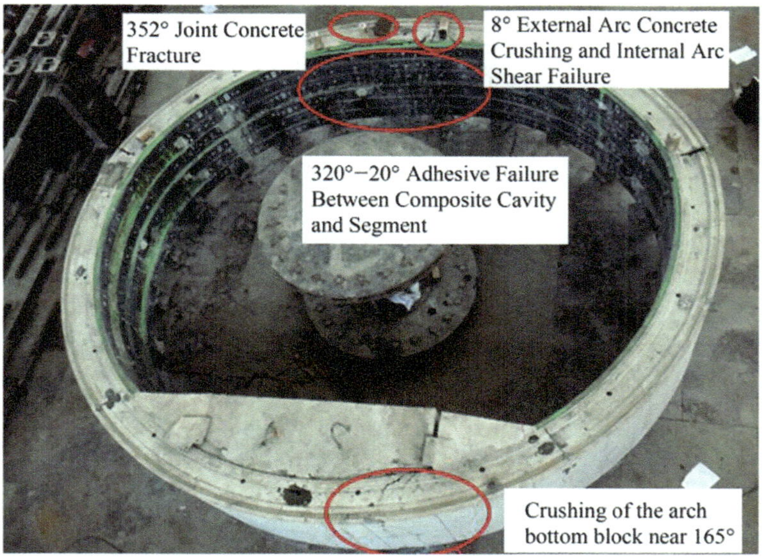

Fig. 4.35 Status after the test of the 12th ring (standard arrangement)

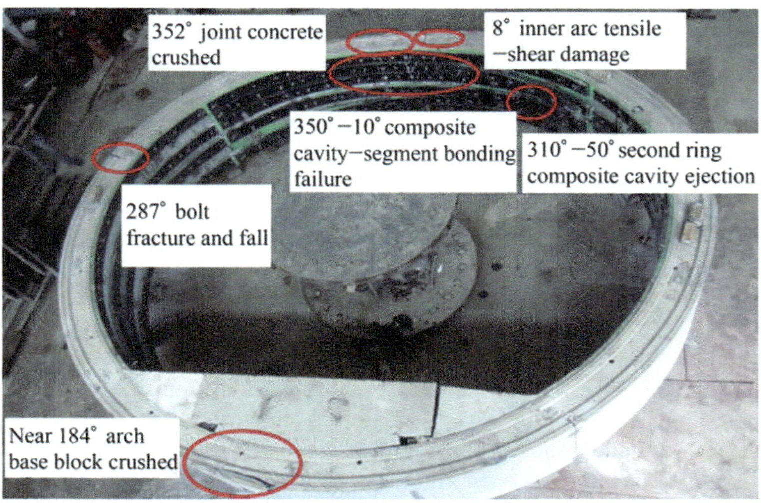

Fig. 4.36 Status after the load–displacement test of 13th ring (reinforced layout)

4.3.6 *Impact of Loading and Unloading Methods*

The study's full-scale test is categorized into two conditions: ground overload condition and lateral unloading condition. The overload condition leads to an increase in the load at the tunnel top caused by soil accumulation or new construction activities. As depicted in Fig. 4.12, Rings 1, 2, 4, 5, 8, 10, 11, 12, and 13 are all subjected to the overload method. Conversely, the unloading condition simulates lateral unloading of an existing shield tunnel caused by activities such as excavation of ultra-deep foundation pits. The specific loading technique is illustrated in Fig. 4.13. Rings 3, 6, 7, and 9 are subjected using the unloading method. Notably, Rings 3, 6, and 7 are paired with Rings 2, 4, and 5, respectively, demonstrating the impact of different loading and unloading conditions on the tunnel lining structure.

Comparing the test phenomena and the load–displacement curves, namely Figs. 4.21 and 4.22, the main conclusions are as follows: damage under different conditions is concentrated at the joints; the first local damage of the structure is reflected in the core area of the top joint, where the concrete is crushed; the sequence of joint failure was first the top joint, followed by the springline joint.

The joint damage in overload condition is differ from unloading conditions. In the overload condition, the axial force of the structure is large, and the edge of the concrete is crushed first in the joint. In the unloading condition, bolt yielding occurs first. Compared to the overload condition, the axial force in the unloading condition continuously decreases, making the structural stress mode more disadvantageous and therefore entering the plastic phase earlier. However, it exhibits better ductility, and the structural deformation at the ultimate state is greater than that in the overload condition.

4.4 Numerical Analysis of Tunnel Structure Load-Bearing Performance

A three-dimensional finite element model of the tunnel lining structure was established using ABAQUS. This model was used to calculate and analyze the relationship between the load (vertical load) and deformation (diameter deformation) of the subway shield tunnel lining segment under test load conditions. The internal force progression of the tunnel segments during deformation was also assessed. Moreover, when the diameter deformation of the subway shield tunnel lining segment reaches a certain convergence threshold under the test load conditions, reinforcement with steel plates is implemented. Subsequent calculations are conducted to derive the complete load (vertical load) to deformation (diameter deformation) relationship, as well as the internal force evolution of the tunnel segments as they deform. By comparing these results with actual experimental data, the reliability and validity of the finite element analysis are verified, providing a reference for practical engineering applications.

4.4.1 Numerical Model

A finite element model of the half ring for a specific subway tunnel is depicted in Fig. 4.37 considering the symmetrical characteristic. In accordance with Chinese standards, we adopted the concrete plastic damage model in ABAQUS, with computation parameters presented in Table 4.3. The remaining materials are steel, having a modulus of elasticity (E) of 2×10^5 MPa and a Poisson's ratio (v) of 0.3. The bolts have a yield strength of 400 MPa. During meshing, considering that the joints are vulnerable points of the segments, the mesh is densified near these joints. The model is composed of solid elements for the concrete body and bar elements to simulate the steel reinforcement and bolts, amounting to a total of 3387 elements.

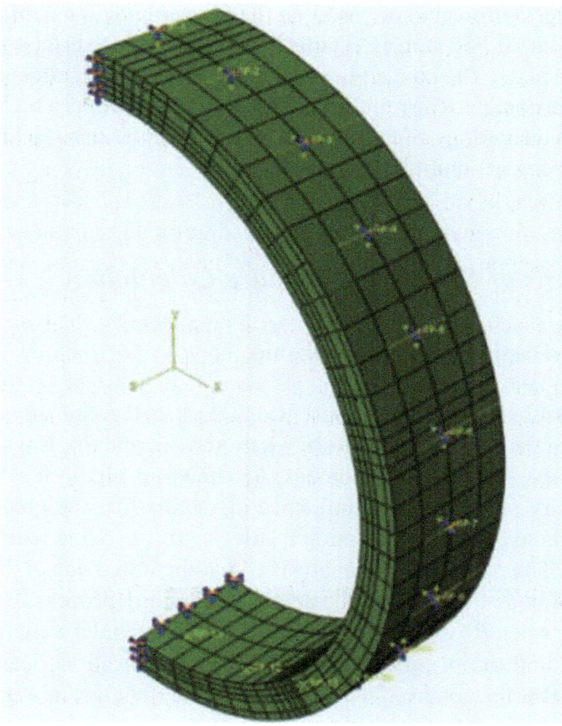

Fig. 4.37 Finite element model

Table 4.3 Calculation parameters for concrete plastic damage model

E (MPa)	υ	ρ (kg mm^{-3})	φ (°)	μ	σ_{cu} (MPa)	c	σ_{tu} (MPa)	σ_{b0}/σ_{c0}	K
335.00	0.2	2.650×10^{-6}	35	0	35.5	0.1	2.74	1.45	0.7

In the finite element model, to avoid the problem of stress singularity caused by applying line loads to solid elements, reference points (RP) are established at the corresponding angles. These reference points are motion-coupled with the nodes on the angle line, establishing a motion constraint relationship between the constrained area and the reference point. Concentrated loads are directly applied to this geometric reference point. The load grouping and proportion relationship are consistent with the experiment. Through 24 concentrated load points, continuous distributed geological resistance, hydrostatic pressure, surface overloading, and other loads are simulated. The equivalent coefficient of load effects can be calculated and adjusted. All horizontal loads in the experiment are divided into three groups: P_1, P_2, and P_3. The ratio P_2/P_1 represents the coefficient of geostatic lateral pressure, which can be selected at different values. The circumferential load grouping is shown in Fig. 4.11.

For the finite element model of the metro tunnel's complete ring with internal tension steel ring reinforcement, based on the above model, a semicircular surface shell with a width of 850 mm is established on the inner surface to simulate the reinforcing steel plate. The contact between the steel plate and concrete is simulated using multi-point contact. The reinforcement calculation adopts a multi-step analysis method, and the technology of birth-and-death elements is used to kill and generate reinforcement elements in different analysis steps.

4.4.2 Analysis of Prototype Structure Calculation

4.4.2.1 Load–Displacement Relationship

The lateral pressure coefficients of test Rings 2, 4, and 5 are the same, all being 0.65. They share the same loading path, where the 4th and 5th rings take the load–displacement curve before reinforcement. As shown in Fig. 4.38, the initial states of the curves vary slightly due to influence of factors like the quality of segment assembly. The finite element calculation results are more in line with the test results of the fifth ring. The reason for the analysis is that the initial gap at the joint leads to the relaxation of bolt stress during the actual test loading process. However, during the 5th ring test, when P_1 was loaded to about 100 kN, all bolts were tightened again, effectively avoiding the phenomenon of bolt stress relaxation. The finite element model assumes that the concrete at the joint is in close contact and ignores gaps and local detachments caused by manufacturing or construction factors, so there is no stress relaxation of bolts during the entire loading process.

4.4.2.2 Lining Structure Stress Analysis

Combining the overall deformation (Fig. 4.39) and the concrete equivalent stress cloud map (Fig. 4.40), it can be concluded that the stress distribution pattern of the lining structure under load is as follows: In the vicinity of the segments at 0 and 180°,

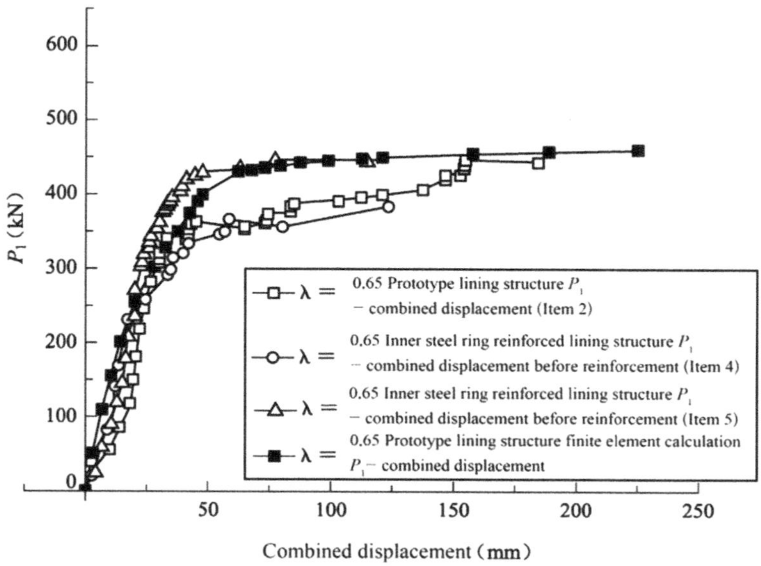

Fig. 4.38 Finite element calculation and test P_1—displacement curve of corresponding points

the inside is under tensile stress, and the outside is under compressive stress. Near the 90° segment, the outside is under tensile stress and the inside is under compressive stress. The external side of the 8° joint, the internal side of the 73° joint, and the external side of the 138° joint all have pronounced areas of compressive stress in the concrete, which precisely corresponds to the areas where the concrete experienced compressive failure in the tests.

As shown in Figs. 4.41, 4.42 and 4.43, during the entire loading process, the reinforcement did not yield, and there was no obvious damage to the main body of the tunnel lining segment. The damage manifested in the segment joints, with damage modes including: (1) Joint opening and misalignment; (2) Pronounced compressive stress concentration in the joint concrete, leading to compressive failure at the joint; (3) Bolt tensile failure. In the finite element analysis, the stress development process began with the top bolts reaching the yield stress, followed by the middle bolts reaching the yield stress.

4.4.3 Calculation Analysis of the Reinforcement Structure

4.4.3.1 Load–Displacement Relationship

Test Ring 4 and Ring 5 are the ultimate load-bearing tests for the inner tensioned steel ring reinforced lining structure, and they have the same lateral pressure coefficient

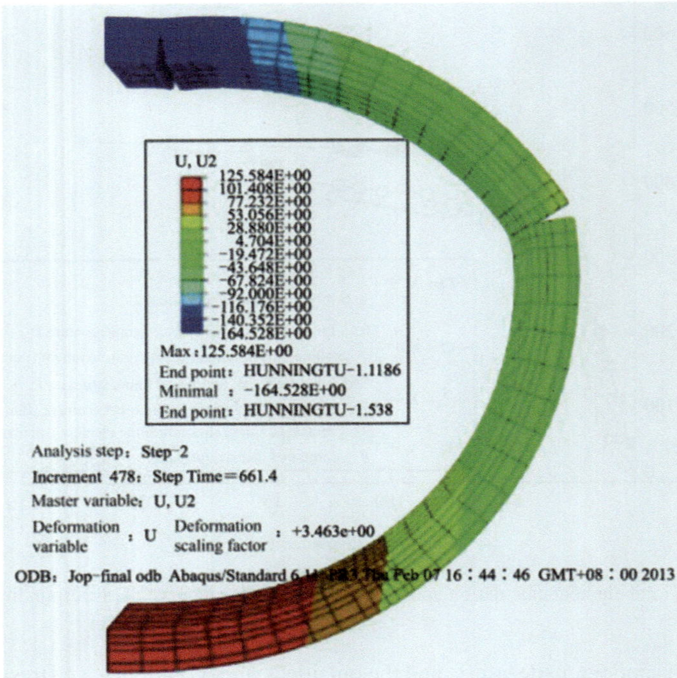

Fig. 4.39 Vertical displacement contour map and deformation pattern

and loading path. Therefore, the results of these two ring tests are compared and analyzed with the finite element calculation results. As can be seen from Fig. 4.44, the overall structural stiffness is greatly improved after using rigid steel plates for reinforcement.

4.4.3.2 Stress Analysis of Lining Structure

The stress cloud diagram of the concrete structure (Fig. 4.45) is basically the same as that of the prototype lining structure. From the steel plate stress diagram (Fig. 4.46), it can be seen that the stress concentration at the anchoring points of the reinforcing steel plate is obvious, especially in the area of the top seal block. Insufficient anchoring is the main cause of damage. This also verifies that during the tests, the failure of bonding and anchoring between the steel plate and the concrete led to a rapid decrease in the overall stiffness of the lining structure, and the entire specimen reached the ultimate bearing capacity.

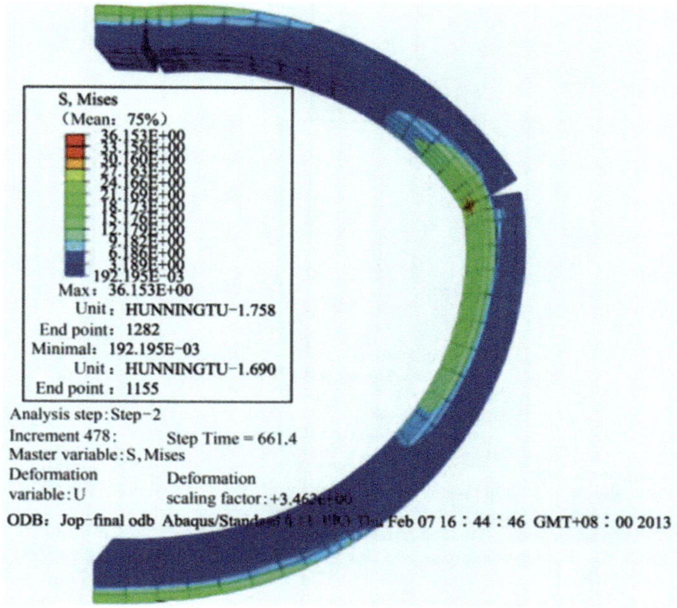

Fig. 4.40 Concrete equivalent stress cloud diagram

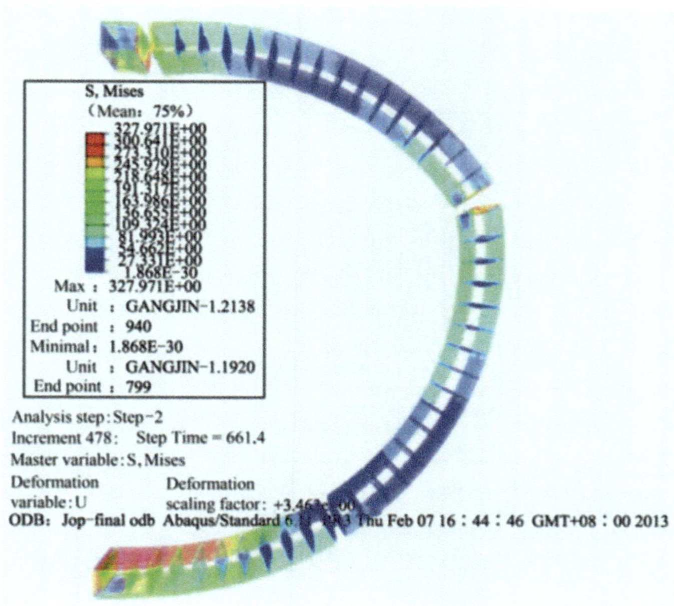

Fig. 4.41 Reinforcement equivalent stress cloud diagram

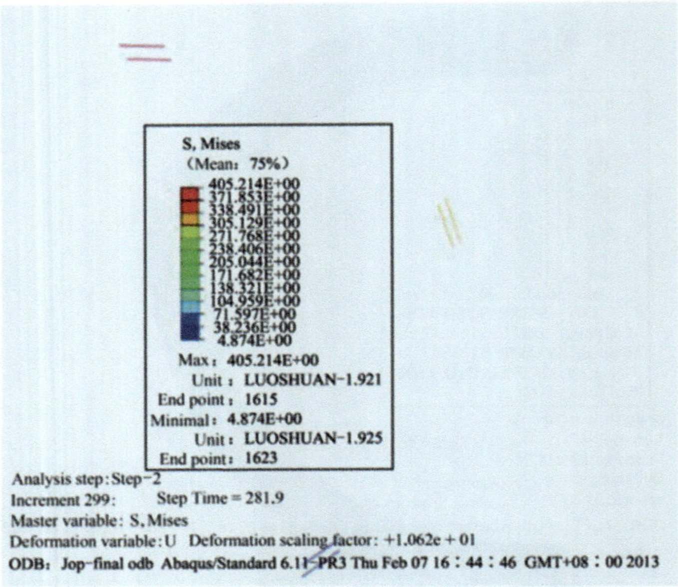

Fig. 4.42 Top bolt reaching yield stress

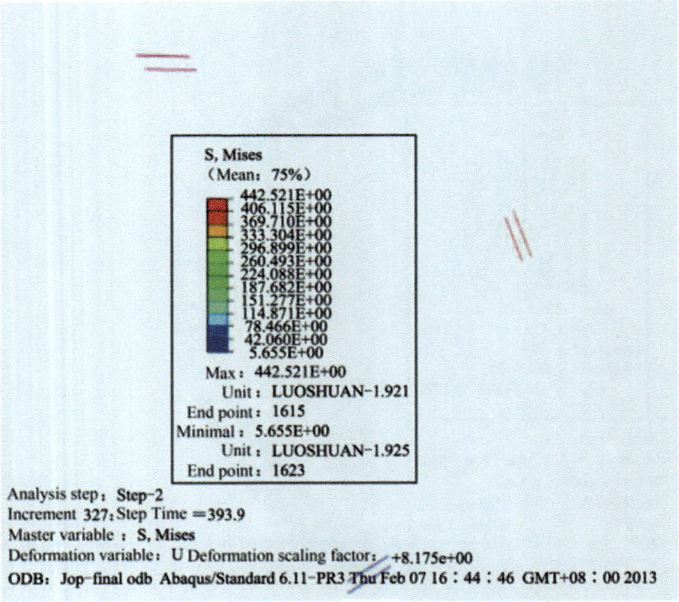

Fig. 4.43 Middle bolt reaching yield stress

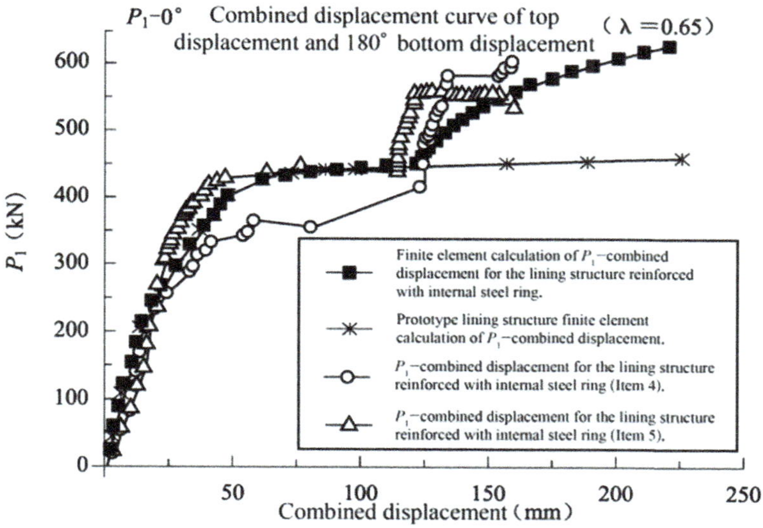

Fig. 4.44 Finite element calculation and experimental P_1-point displacement curve

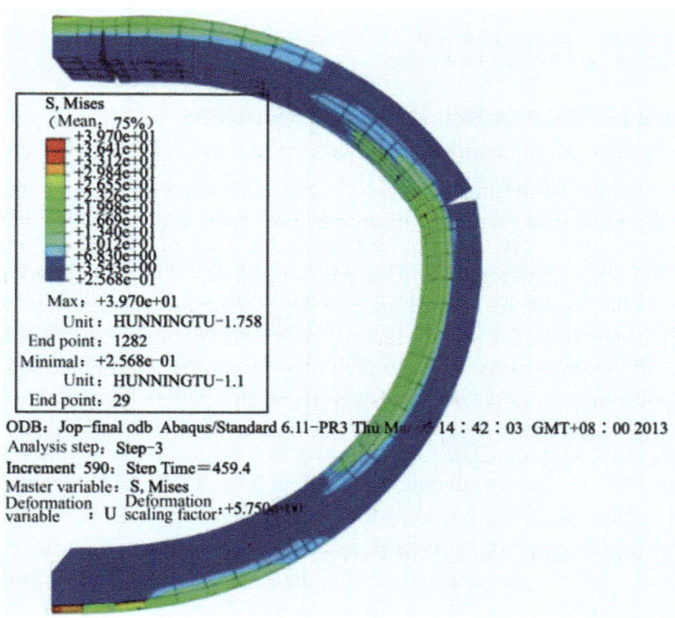

Fig. 4.45 Equivalent stress contour map of concrete

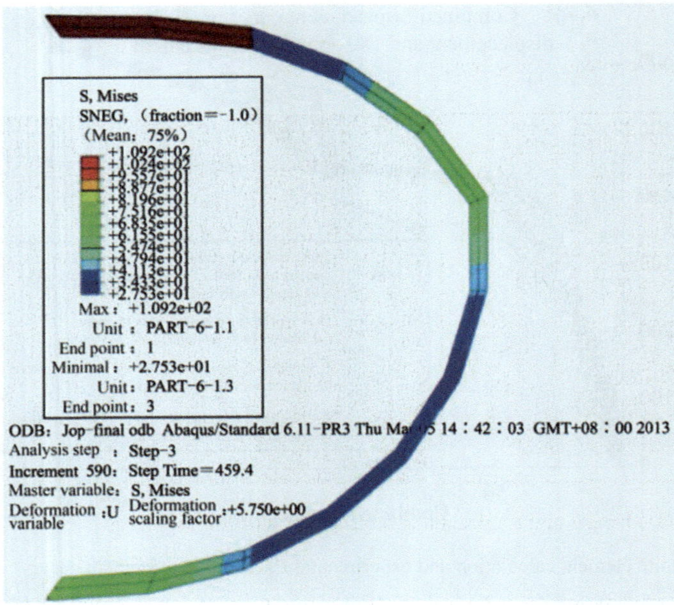

Fig. 4.46 Equivalent stress contour map of reinforced steel plate

4.5 Tunnel Maintenance Recommendations

Based on the results from tunnel structure experiments and theoretical analyses, the following recommendations and opinions are given for tunnel structure maintenance:

(1) For segmental tunnels, as long as the deformation remains within the permissible limits set by the design and construction acceptance standards (tunnel convergence deformation of 2–5‰D), it can meet design specifications and operational requirements. However, when the tunnel convergence deformation approaches or exceeds 20‰D, the tunnel becomes notably sensitive to changes in external loads. This can result in significant joint opening and localized compression damage of the segment. The limit of tunnel deformation (in the joints) is approximately 15 ~ 18 cm, which equates to about 25 ~ 30D‰. The primary form of tunnel damage is characterized by bolt-joint failure. In experiments, the ultimate failure is observed when the concrete on the external side of the joint connecting the F and L blocks gets crushed, leading to a considerable decline in the overall structural rigidity.

(2) Segmental tunnels are more sensitive to deformation under external lateral unloading conditions than under the conditions of upper overload. This has a significant impact on the load-bearing capacity of shield tunnels. When the tunnel is in a state of force balance, the influence of lateral unloading activities around the tunnel structure should be strictly controlled.

(3) If the convergence deformation of the segmental tunnel exceeds 20‰D, it is imperative to promptly reinforce the tunnel structure. Through experimental and theoretical analysis, it is suggested to control the significant deformation of segmental tunnels by combining both internal and external methods. Internally, the use of tensioned steel rings and steel–concrete composite structures can be adopted as repair techniques. After the repair, they can both meet the normal usage requirements of the tunnel.
(4) Using internal steel rings to reinforce the segmental structure during later stages can effectively enhance the tunnel structure's overall bending stiffness and ultimate load-bearing capacity. Force characteristic analysis of the two through experiments revealed that the failure mode of the reinforced tunnel is adhesive-solidification failure. Although the load-bearing capacity of the tunnel is improved after reinforcement, its adaptability is relatively weaker in case of significant deformation in the later stages.

4.6 Summary

In this section, a full-scale test was conducted on the unreinforced shield tunnel lining over a span of 7 rings. The test considered various parameters including different lateral pressure coefficients, the condition of the bolts, loading and unloading scenarios, reinforcement, and the ultimate load-bearing capacity. Upon in-depth analysis of the test data, the primary conclusions are:

(1) The lateral pressure coefficient exerts a significant influence on the mechanical behavior of the tunnel lining. The conditions involving peripheral unloading prove to be more perilous than those with surface overloading. This insight offers valuable guidance for deformation control during pit excavation.
(2) The mechanical response of the bolted segment lining can be delineated into three distinct phases: elastic, elastoplastic, and plastic. The unbolted lining structure, in contrast, lacks a pronounced elastoplastic phase.
(3) During the structure's elastic phase, corrosion of the bolts has minimal impact on its overall stiffness. However, failure of the bolts results in a noticeable decline in overall stiffness in the elastic phase. In the elastoplastic phase, the presence of circumferential bolts affects the stiffness. During the plastic phase, corrosion and failure of the bolts lead to a reduction in the structure's ultimate load-bearing capacity.
(4) The segmental joints play a pivotal role in governing the mechanical behavior of the tunnel lining. During structural failure, plastic hinges are formed at the locations of the four joints, making the lining structure undergo geometric variation.
(5) To achieve failure in a bolted lining structure, plastic hinges must manifest at both the top and springline joints. For unbolted structures, plastic deformation is only required at the top joint.

(6) Based on the structural performance metrics, the shield tunneling method used in Shanghai can be categorized into five distinct performance levels.

A full-scale test spanning 4 rings was conducted on the steel-reinforced shield tunnel lining structure, considering different reinforcement scopes and both overloading and unloading scenarios. After an in-depth analysis of the test data, the main conclusions include:

(1) Steel plate reinforcement effectively enhances the load-bearing capacity and stiffness of the tunnel lining ring.
(2) The reinforced structure undergoes brittle failure, primarily characterized by adhesive failure at the top and bottom.
(3) Increasing the width of the steel plate at the top did not alter the failure mode of the reinforced structure. Whole-ring reinforcement, compared to half-ring reinforcement, exhibits greater integrity and slightly better performance.
(4) From a testing perspective, eliminating the corbel structure and reducing the bottom reinforcement scope can increase construction efficiency without significantly impacting the ultimate load-bearing capacity. However, it does diminish the overall stiffness of the lining ring. If the corbel structure is omitted on-site, both actual construction efficiency (due to a lack of support) and safety would considerably decrease (implementation of tension bars also becomes challenging).
(5) The use of corbels and tension bars enhances the structural load distribution and increases its load-bearing capacity.
(6) Both the test and on-site construction indicate that the components—corbels, steel plates, tension bars, bolts, epoxy—must be reliably bonded (connected) to the segment for a substantial increase in structural load-bearing capacity.

For the novel structural system of reinforced shield tunnel lining with composite cavities, a full-scale test spanning 2 rings was conducted, considering different arrangements of the composite cavities. After thorough data analysis, the main conclusions include:

(1) In terms of reinforcement efficacy, the composite cavity reinforcement is largely comparable to steel plate reinforcement.
(2) From an assembly perspective, reinforcing with composite cavities offers lightweight and convenient characteristics.
(3) Lengthening a portion of the cavity at the top of the tunnel (126°) changes the sequence of joint failure and adhesive failure at the top.
(4) In terms of on-site construction efficiency, the current lack of specialized construction equipment results in very low efficiency.
(5) To enhance the load-bearing capacity of the whole tunnel lining, it's necessary to connect with other structural components to form a complete ring, creating an effective load-bearing system.
(6) Analyzing from the perspective of auxiliary waterproofing effects, steel rings have a distinct advantage over composite cavities.

Reference

Liu X, Tang M, Lu L, Wang X, Zhu Y (2013) Experimental study on the structural bearing capacity of shield tunnel reinforced by internal steel rings—full ring reinforcement method. Chin J Rock Mech Eng 32(11):2300–2306

Chapter 5
Deformation-Induced Tunnel Damage and Repair Techniques for Tunnel Linings

5.1 Introduction

In recent years, with the rapid development of urban rail transit networks across the country and the intensive development of urban underground spaces along metro lines, there have been multiple incidents of significant convergence and settlement deformations in subway tunnels. These incidents affect the safety of tunnel structures and the normal operation of trains. Large deformations in tunnels often lead to a series of significant issues, primarily manifesting as the opening of longitudinal joints on both sides of concrete segment crown blocks (through joints), through-cracks at the top (staggered joints), water seepage at both springlines, and the separation and opening of lining rings and longitudinal joints. Such deformations may even result in damage to the track bed or track misalignment, often accompanied by damage and fragmentation of the concrete segments. Over time, this damage tends to worsen, eventually leading to the instability and failure of the tunnel structures, posing a risk to subway operational safety. Therefore, technical measures are essential for waterproofing and the comprehensive reinforcement of the damaged tunnels.

Both domestically and internationally, common reinforcement methods include enlarging the thickness of concrete segment or applying fiber composite sheets for reinforcement. However, these methods have their inevitable shortcomings. The enlargement of the segment thickness significantly occupies space, making it challenging to implement in the confined spaces of underground tunnels, particularly in operational tunnels with complex boundary requirements. The application of fiber composite materials is relatively straightforward and cost-effective, but these materials have inferior stiffness and almost no resistance to compression or shearing. They also lack waterproofing capabilities. Tests have shown that using carbon fibers and aramid cloth for reinforcing segmental lining does not significantly enhance load-bearing capacity. If not adhered securely, these materials can pose safety risks

R. Wang et al., *Structural Safety Early Warning and Treatment Technology of Metro Tunnel in Soft Soils*, Key Technologies for Tunnel Construction under Complex Geological and Environmental Conditions,
https://doi.org/10.1007/978-981-96-6410-8_5

to trains in operation. Instances abroad have seen the detachment of carbon fibers from concrete structures affecting the operation of high-speed trains, and a similar case occurred in Shanghai in 2009.

Given the numerous challenges faced during structural reinforcement work in operational tunnels, such as "limited time, confined space, and strict boundary requirements," this chapter primarily introduces two methods suitable for repairing and reinforcing tunnel lining in operational tunnels. These are the internal tension steel ring method and the composite cavity construction reinforcement method. They enable a robust combination of "segmental lining, epoxy, bolts, steel rings (or composite cavities)," collectively bearing stress, thereby significantly enhancing structural rigidity.

5.2 Internal Tension Steel Ring Reinforcement Technique

The internal tension steel ring lining reinforcement technology for deformed tunnels is a comprehensive procedure, encompassing several critical steps: initially addressing water leakage areas, reinforcing dislodged track beds, repairing damaged segments, strengthening cracks, stabilizing the deformed structure, and ultimately installing internal steel linings to create an enhanced, reinforced structural system. This method operates on the principle of using circular steel plate supports and externally injecting epoxy resin around the tunnel segments for reinforcement. By integrating internal strengthening with external support, this approach substantially improves and augments the overall integrity and safety of the tunnel structure, thereby significantly enhancing its strength and stiffness, which is essential for the tunnel's prolonged and secure operation.

5.2.1 Principles and Features of the Technique

The principle of the internal tension steel ring reinforcement technique is based on the existing reinforced concrete structure, with the aid of a specialized manipulator (as shown in Fig. 5.1) to install a 2–3 cm thick steel plate along the intrados of the tunnel lining, creating a steel support ring. This steel ring is secured to the concrete segments with approximately 100 bolts, and the gaps are filled with rigid epoxy grout, forming a unified and cohesive circular load-bearing structure. A key feature of the internal tension steel ring reinforcement process is its ability to ensure uninterrupted operation of the tunnel. This technique is designed to be implemented during limited subway downtime (often called window time), using the existing rail and other conditions within the tunnel. It doesn't alter any operational limitations or requirements, such as electrical power needs for operation. The entire reinforcement process, including all auxiliary steps, is carried out using specially designed and

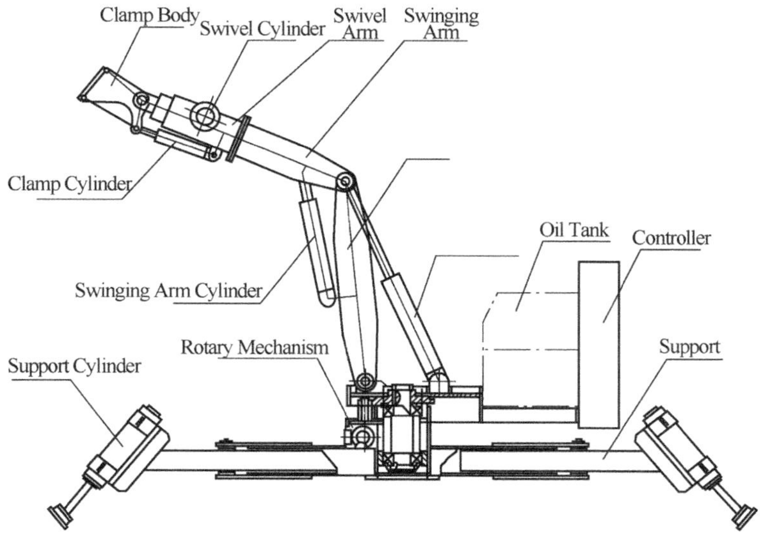

Fig. 5.1 Schematic diagram of manipulator components

fabricated equipment that allows for the installation of the steel ring support without affecting the tunnel's regular operations.

The internal tension steel ring reinforcement technique can be extensively used to manage large transverse convergence deformations and differential longitudinal settlements in tunnels caused by various factors. It is applicable to both operational tunnels and tunnels under construction. This technique can effectively curb the worsening trend of deformation, thereby improving the load-bearing capacity of damaged tunnels. Whether to adopt joint-riding or inner-ring steel hoop installation depends on the tunnel damage situation and construction conditions. Typical reinforcement scenarios and corresponding measures for both under-construction and operational tunnels are shown in Table 5.1.

5.2.2 Design of Internal Steel Hoop

5.2.2.1 Design Parameters

The design of the internal tensioning ring involves several critical aspects such as the form it takes, its placement, segmentation, material selection, and the thickness of the steel plates used.

Table 5.1 Summary of reinforcement situations for damaged tunnel segments

Damage cause	Form of damage	Main measures taken	
		Under-construction tunnel	Operational tunnel
Adjacent construction to tunnels	Transverse convergence	Steel ring installation (full ring plate)	Steel ring installation (with corbel bracket plus ring plate, retaining the track bed)
	Differential settlement	Riding joint steel ring installation (full ring plate)	Riding joint steel ring installation (with corbel bracket plus ring plate)
Overburden overload above the tunnel	Transverse convergence	Steel ring installation (full ring plate)	Steel ring installation (with corbel bracket plus ring plate, retaining the track bed) or steel ring installation (full ring plate, track bed removed)
Direct damage to the tunnel due to other construction accidents	Tunnel segment penetration	–	Continuous steel ring installation (with corbel bracket plus ring plate)
Freezing (pipe) construction of bypass passage	Damage to the main reinforcement of the tunnel segment	Steel ring installation (full ring plate)	–

(1) Form of the Internal Tension Ring

To facilitate processing and construction within the tunnel, the internal tension ring is typically fashioned from steel plates of equal thickness throughout the entire tunnel. The strength of these steel plates should exceed that of grade Q235B.

(2) Width of the Internal Tension Ring

The reinforcement construction of the internal tension ring should complement the original tunnel segment structure. Generally, the ring's width is slightly narrower than that of the tunnel segments for ease of installation and to accommodate the setup of cable brackets within the tunnel. When the tunnel has sustained significant damage, it may be necessary to lay continuous steel rings for several segments, depending on the severity of the damage.

(3) Segmentation of the Internal Tension Ring

Segmentation within the tunnel varies depending on whether it is a construction tunnel or an operational tunnel. In construction tunnels, where transportation conditions are poor because no rail can be used, installation is carried out using hoists, and mechanization is low, segments can be made smaller. In contrast, operational tunnels have better transportation conditions that allow for the transport of larger equipment and steel components via rail, enabling larger segments due to higher levels of automation. The main factors to consider for steel plate reinforcement in operational tunnels depend on the lifting capacity of the installation equipment inside the tunnel, the processing capability of the factory, and the convenience of the construction site.

During construction within the tunnel, the weight and forces on the internal tension ring should satisfy the following considerations:

a. Ease of transport. This includes the entire transportation process from the reinforcement stage to the installation stage, accounting for reinforcement equipment, transportation devices, transshipment, storage, and lifting.
b. Ease of installation. The installation phase should be convenient and quick to allow for rapid construction.
c. Pipeline avoidance. Large segments can easily damage essential pipelines inside the tunnel, such as high-voltage lines and signal lines. Weld joints should avoid these main pipelines to prevent damage during the welding process.
d. Segmentation should avoid the longitudinal joints of the segments, and weld joints should be at least 30 cm away from the joints to ensure better rigidity.

Taking all these factors into account, the segments are welded together, aiming to make each segment as large as possible and the number of segments as few as possible. Typically, a full-circle steel support is divided into 4 to 6 steel plates, with two corbel plates, two side plates, and one top sealing plate, with the heaviest steel plate weighing up to 500 kg. The segmentation of the steel plate is shown in Fig. 5.2.

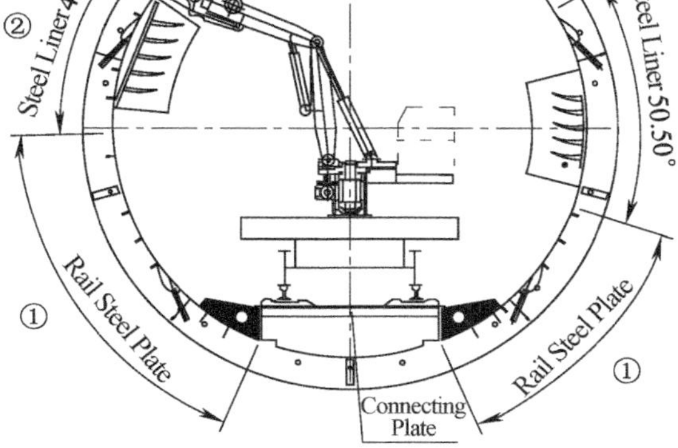

Fig. 5.2 Segmentation diagram for internal tension ring reinforcement. *Note* Installation Procedure 1-2-3-4-5

For tunnels under construction and operational tunnels, different forms are chosen respectively. For tunnels under construction, the entire ring form, which offers better load-bearing capacity, is used. In operational tunnels, due to the existing track bed, it's necessary to adjust the bottom connection method. Steel corbels are set on both sides of the track bed, and steel tie rods are used to connect the steel corbels, replacing the bottom steel ring. This enhances the overall integrity of the steel liner, enabling better overall load distribution, increased stiffness, and greater reliability in reinforcement.

5.2.2.2 Design Drawings

The steel lining of a full ring consists of six or more steel plates which, after positioning and installation, are welded together to form the complete ring. For steel rings where the track bed is cut and corbels are set up, the assembly generally includes five steel plates. These plates are welded together, and steel bars are welded between the corbels to connect them. The segmentation and cross-sections of the internal steel ring are illustrated in Figs. 5.3, 5.4 and 5.5.

Design details of inner steel rings at various locations as shown in Figs. 5.6, 5.7, 5.8 and 5.9.

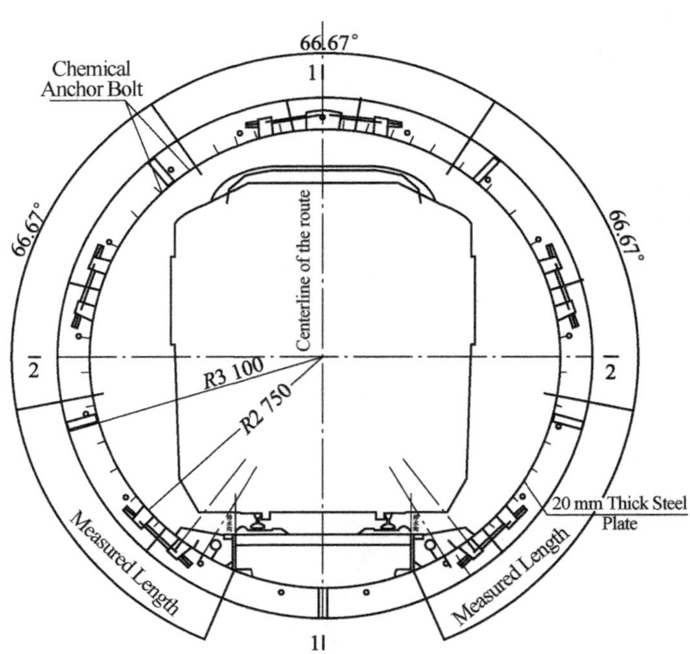

Fig. 5.3 Segment diagram of internal steel ring reinforcement

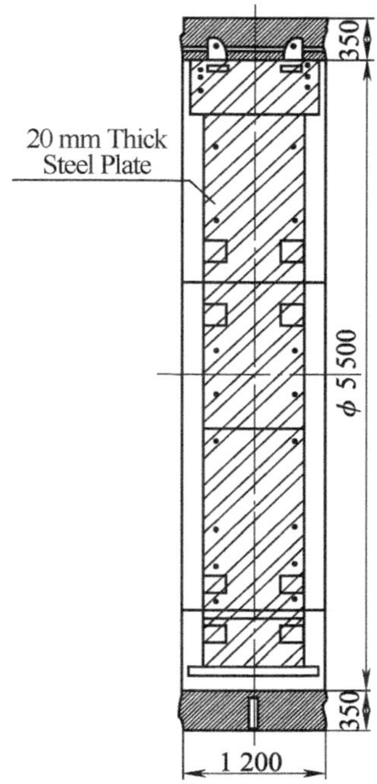

Fig. 5.4 Section 1-1 of inner steel ring (unit: mm)

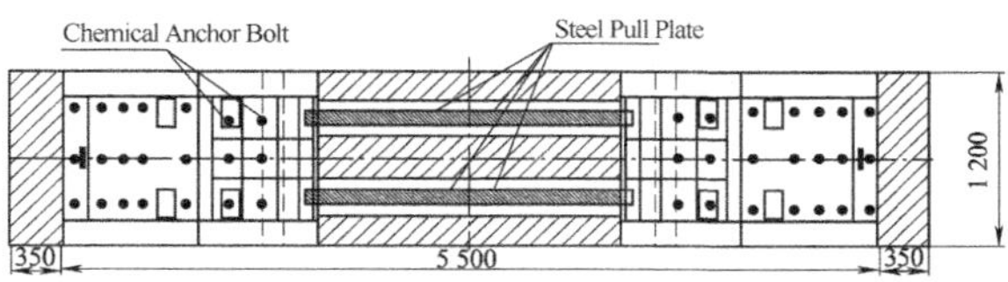

Fig. 5.5 Section 2-2 of inner steel ring (unit: mm)

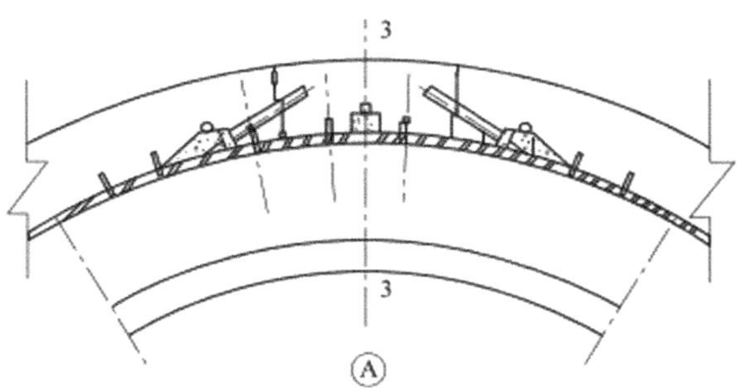

Fig. 5.6 Detailed installation of inner steel ring at the top of the tunnel

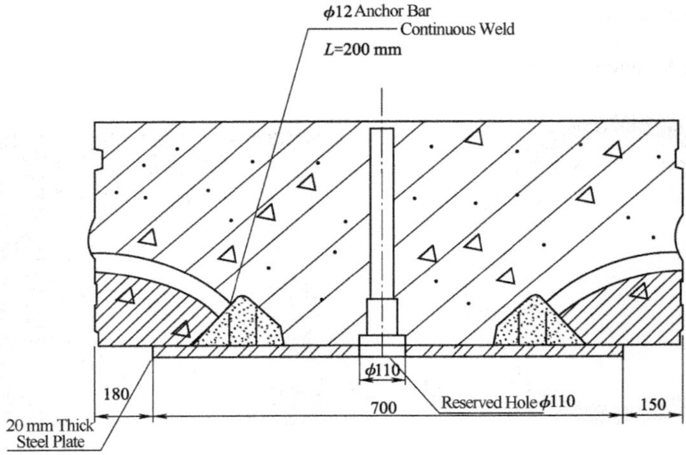

Fig. 5.7 Detailed sectional view 3-3 of inner steel ring installation at the top of the tunnel (unit: mm)

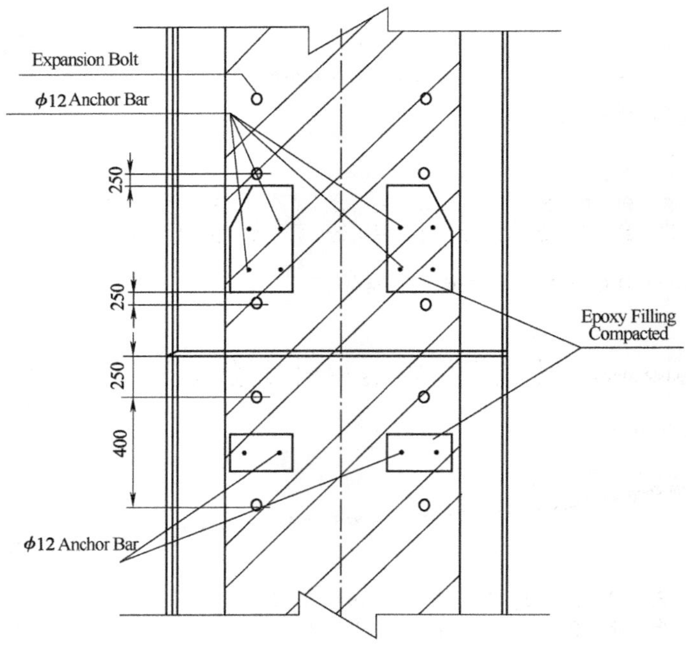

Fig. 5.8 Detailed schematic of steel ring installation at tunnel side joint (unit: mm)

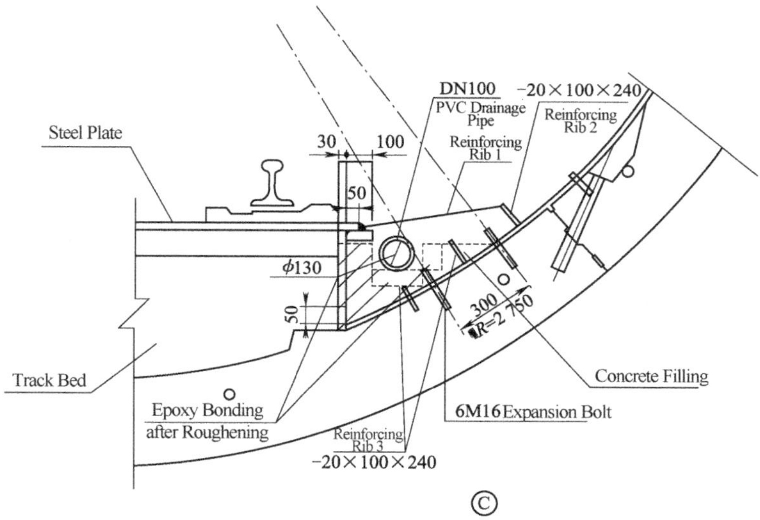

Fig. 5.9 Schematic detail of steel calf corbel structure and installation at the tunnel bottom sleeper location (unit: mm)

5.2.3 Mechanical Equipment for Reinforcement

To meet the requirements for transportation and operational states, the materials, mechanical equipment, and work platforms used must not exceed the operational limits of the tunnel. They should not interfere with or damage any cables or pipelines within the tunnel. Specifically for operational tunnels, the work must be completed within the scheduled shutdown window time to avoid disrupting the overall operation of the tunnel and affecting subsequent operations. Hence, the design and use of equipment have high demands in terms of spatial constraints and construction efficiency.

Taking the tunnel with an inner diameter of Φ5.5 m in Shanghai as an example, the internal clearance conditions of the tunnel is shown in Fig. 5.10. To ensure work efficiency, improve construction accuracy and safety, a special equipment platform (or construction robot) specifically designed for tunnel steel ring reinforcement was developed. In addition to being able to lift a certain weight, it is also required to make multiple degrees of adjustment within the specified space to complete actions such as gripping, flipping, passing through pipelines, and positioning steel plates. The integrated specialized machinery is shown in Fig. 5.11, which includes a gripper mechanism (mechanical arm), two actuator arms, and a movable construction platform.

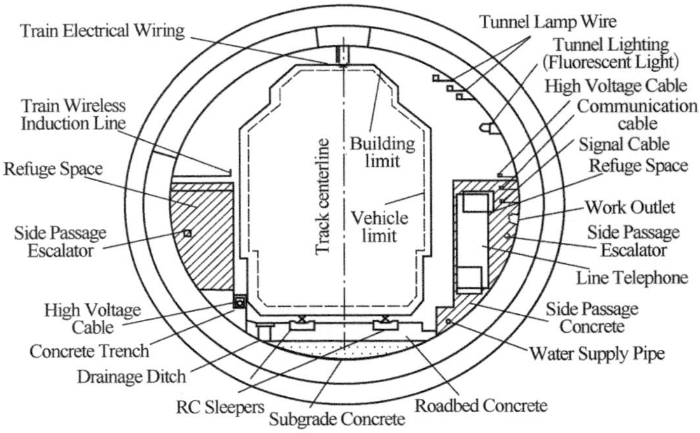

Fig. 5.10 Schematic diagram of the interior limitations of the tunnel

Fig. 5.11 Schematic of the mechanical arm

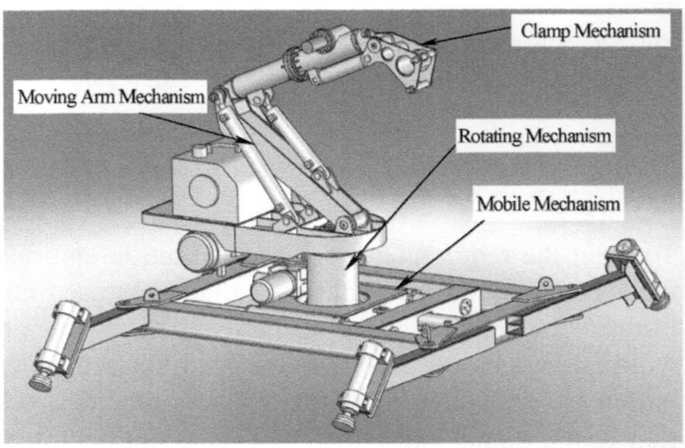

5.2.3.1 Gripper Mechanism (Mechanical Arm)

The mechanical arm is mainly controlled by electrical components to operate its various mechanical parts. The operational power for each component is provided by an integrated hydraulic system. The mechanical arm can autonomously pick up steel liners from the platform, complete the assembly actions required for the installation of the steel liners, and accurately position them on the inner wall of the tunnel lining. Within a Φ5.5 m tunnel space, it can handle an arched steel liner with an arc length of 3250 mm, a width of 700 mm, a thickness of 20 mm, and a radius of 2750 mm, and navigate it through tunnel pipelines to assemble it on the inner wall of the Φ5.5 m tunnel. Its main systems consist of a mobility mechanism, a rotating mechanism, an arm mechanism, and a gripper mechanism, as shown in Fig. 5.12.

Fig. 5.12 Photo of the mechanical arm

In the use of the mechanical arm, the control of its degrees of freedom is a crucial aspect. To ensure precise installation of the steel liner and to meet the functional requirements for steel liner assembly and construction, the mechanical arm is equipped with the following degrees of freedom:

(1) Arm Extension: Mainly used to position the steel liner accurately.
(2) Swing Arm Extension: Primarily facilitates the accurate positioning of the steel liner.
(3) Clamp Extension: Used mainly for adjusting the angle of the curved steel liner, ensuring it aligns with the tunnel lining and passes through conduits correctly.
(4) Clamp Rotation Around the Arm Axis: Essential for adjusting the angle of the curved steel liner, ensuring precise alignment between the steel liner's curved surface and the tunnel lining.
(5) Rotation: Primarily for rotating the mechanical arm horizontally from the point where it picks up the steel liner to the assembly position.
(6) Translation: Used for adjusting the longitudinal position of the steel liner along the tunnel, ensuring that the steel liner can be installed at the center of the tunnel segment.

5.2.3.2 Swing Arm Crane

Due to the maximum weight of a single steel liner being 500 kg, electric swing arm cranes are essential for the transportation and installation of the steel liners and track plates. Currently, two electric swing arm cranes with a lifting capacity of 1 ton are used.

5.2.3.3 Construction Platform

As the steel liners are installed on the inner walls of a Φ5.5 m tunnel, a construction platform is required to install the expansion bolts for securing the steel liners and for spot welding and welding the segmented steel liners. Therefore, two types of construction platforms have been designed: one mounted on a flatbed vehicle and the other is a detachable platform.

The operating platform mounted on the flatbed vehicle can move along it, coordinating with the mechanical arm to install the steel liners. To ensure the construction platform does not exceed the metro tunnel's internal clearance, a specially designed two-layer foldable and retractable footplate is utilized to meet operational needs. The other detachable construction platform, assembled and disassembled manually, is used specifically for installing expansion bolts for the steel liners and for welding operations.

5.2.4 Construction Process and Key Operation Points

For tunnel structures requiring internal steel ring reinforcement, other defects are often present. Therefore, while carrying out the reinforcement, it's common to first address these defects. The main steps of the internal steel ring reinforcement process include: sealing leakage points, reinforcing detached track beds, repairing damaged segments, strengthening cracks, stabilizing deformations, and finally installing the steel lining to form a reinforced new load-bearing system. The specific construction process is illustrated in Fig. 5.13.

Structural Waterproofing: The waterproofing strategy utilizes the inherent waterproof characteristics of the original reinforced concrete structure and the elastic sealing gaskets of the segments. The soil around the segments is improved by pressure grouting with materials such as polyurethane slurry, while epoxy grout is injected into the circumferential and longitudinal joints, creating multiple layers of waterproofing lines to ensure effective sealing of water leakage.

Structural Reinforcement: Structural reinforcement is based on the original reinforced concrete structure, supplemented by installing a 20–30 mm thick steel plate ring along the inner arc of the segments, creating a steel ring support. For construction tunnels, full-circle steel plates are used; for operational tunnels, due to the presence of the track bed, additional steel corbels and tie bars are used atop the ring plates to create an integrated structure. The steel ring and the concrete segments are unified by filling the gaps with rigid epoxy grout, forming a new circular load-bearing structure.

Emergency Treatment: Considering the needs of emergency situations, structural reinforcement typically involves two aspects: firstly, to prevent further deformation of the tunnel, temporary reinforcement involves applying aramid fabric to the tunnel surface; secondly, permanent reinforcement involves installing circular steel rings inside specific segments to provide lasting structural support.

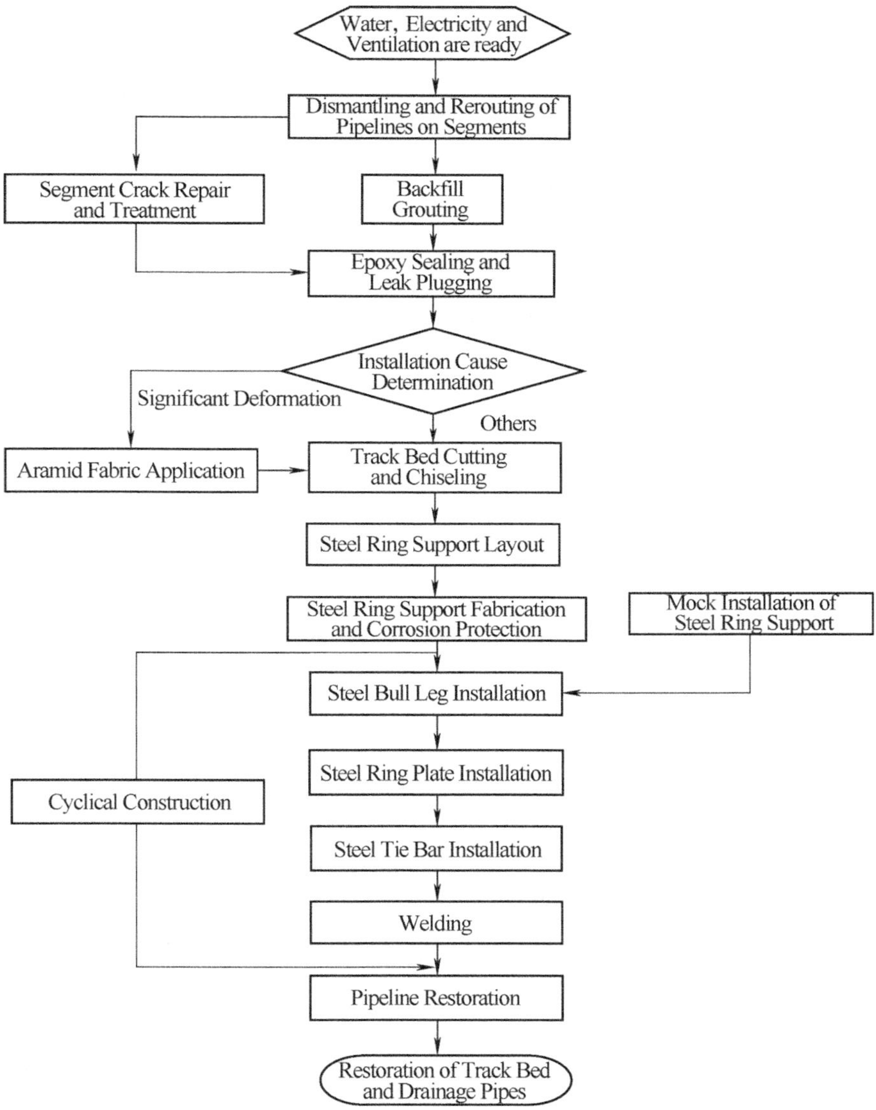

Fig. 5.13 Detailed flowchart of steel ring support construction in operational tunnels

5.2.4.1 Template Setting for the Internal Steel Rings

Before construction, each internal steel ring should be templated using a plastic board. Based on the templated dimensions, the internal steel rings are fabricated in the factory to meet the reinforcement design specifications. Especially when the

Fig. 5.14 Construction process of internal steel ring dimension layout

reinforcement section is located in a section with a small turning radius, and there is a significant tilt in the track bed, the cross-sectional dimensions of the steel track plate on both sides differ significantly from the design drawings, and each ring varies. Therefore, it is necessary to process and manufacture the steel track plates on both sides according to the templates set at the reinforcement site. Additionally, due to the deformation of the tunnel cross-section, the actual curvature and arc length of the upper steel liner differ from the theoretical values, and thus the steel liner shape and arc length need to be adjusted based on the templates. On-site templating is shown in Fig. 5.14. Nowadays, 3D laser scanning or total station equipment is also used on-site, which can accurately depict the misalignments and deformations of the tunnel inner walls. Digitization facilitates the partitioning and processing of the segments.

5.2.4.2 Internal Steel Ring Manufacturing

Based on the design drawings, the internal steel lining is manufactured by a specialized unit according to the on-site layout dimensions of the internal steel ring. The processed internal steel rings are numbered as left steel track plate, left lining plate, top lining plate, right top lining plate, right bottom lining plate, right steel track plate, etc. Holes for expansion bolts are pre-drilled, and the surface is coated with black epoxy paint.

Steel track plates and steel lining plates are manufactured in the factory according to the layout and then transported to the site for installation.

5.2.4.3 Manufacturing and Installation of Connecting Plates

The specific dimensions of the connecting plates for the top of the track bed are determined based on the on-site layout results and manufactured in the factory according to the design requirements. The connecting plates, used to connect the left and right steel track plates, are welded at both ends to the steel track plates. Since the whole connecting plate cannot pass under the subway tracks, it is divided into two parts and welded on-site.

5.2.4.4 Preliminary Preparations for Internal Steel Ring Installation

Prior to installation, the following preparations should be made:

(1) Cleaning of the segment: High-pressure water jets are used to clean the segment.
(2) Application of epoxy primer: To ensure full adhesion of the epoxy filler to the segment, epoxy primer is applied to the cleaned segment.
(3) Cutting of the track bed: The original track bed needs to be cut to create space for the installation of the steel track plates.
(4) Material preparation: The materials for the internal steel ring installation are transported to the construction site by flatbed trucks. The quantity of materials transported each time should be adjusted according to construction needs due to the time constraints imposed by subway operations.

5.2.4.5 Installation of Corbel Braces

The construction sequence for installing corbel braces on both sides of the track bed is as follows:

(1) Remove some of the concrete debris from the sides of the track bed to ensure sufficient construction space.
(2) Roughen the concrete surface on the sides of the track bed and the inner arc of the segment in advance to ensure strong epoxy adhesion.
(3) Connect the corbel brace ring plate to the inner arc of the segment using six M16 expansion bolts (arranged along the inner arc of the segment, spaced 300 mm apart and embedded 100 mm deep).
(4) Considering the drainage requirements of the tunnel, PVC drainage pipes are installed inside the corbel brace.
(5) The corbel brace is designed with injection holes. After all ring plates and track bed steel plates are installed, use these holes to fill the space between the brace, the segment, and the track bed with epoxy resin.

The structure of the corbel brace is illustrated in Figs. 5.15, 5.16, 5.17 and 5.18.

5.2.4.6 Installation of the Internal Steel Ring

The mechanical arm sequentially installs the segments of the internal steel lining from bottom to top symmetrically. During construction, the center of the mechanical arm is aligned with the location where the steel lining segment is to be installed,

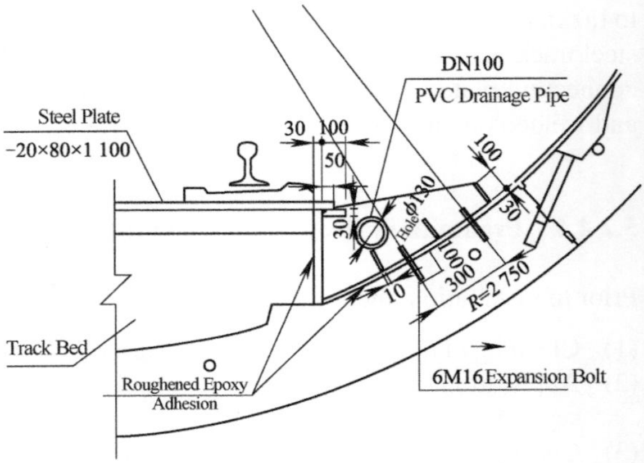

Fig. 5.15 Schematic of corbel brace structure (unit: mm)

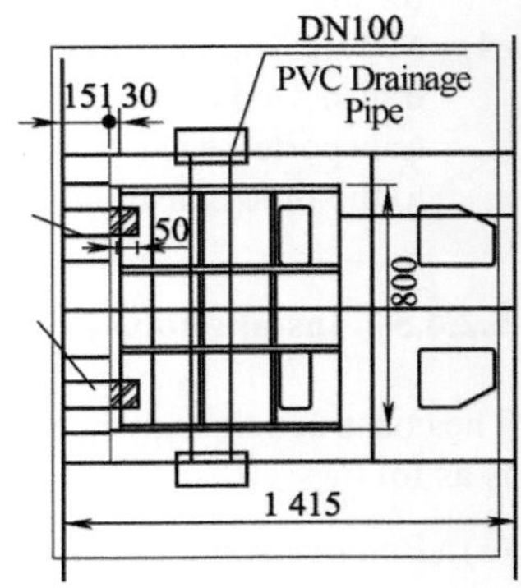

Fig. 5.16 Schematic plan of corbel support structure (unit: mm)

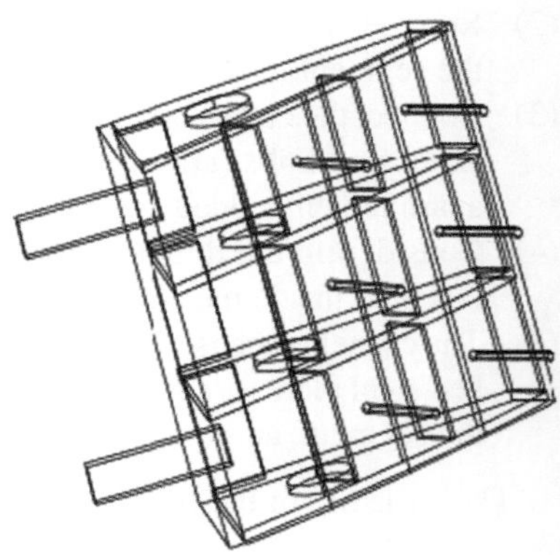

Fig. 5.17 Schematic of corbel support structure

Fig. 5.18 Schematic of the installation process of corbel support structure

Fig. 5.19 Installation process of the internal steel ring 1

and the support is then extended. The mechanical arm continues to symmetrically install other steel lining segments from bottom to top. After each segment of the steel lining is positioned, it is temporarily fixed with expansion bolts and spot-welded to the adjacent steel lining. The joints between the steel plates are then welded together. The process of installing the internal steel ring is illustrated in Figs. 5.19, 5.20, 5.21, and 5.22.

5.2.4.7 Installation of Steel Plates on the Track Bed

Taking into account potential dimensional errors during the installation of the ring plates, the connection of steel plates between corbel brackets on the track bed is carried out after all ring plates are installed. The steel plates are placed between the sleepers. The process of installing steel plates on the track bed is shown in Fig. 5.23.

Fig. 5.20 Installation process of the internal steel ring 2

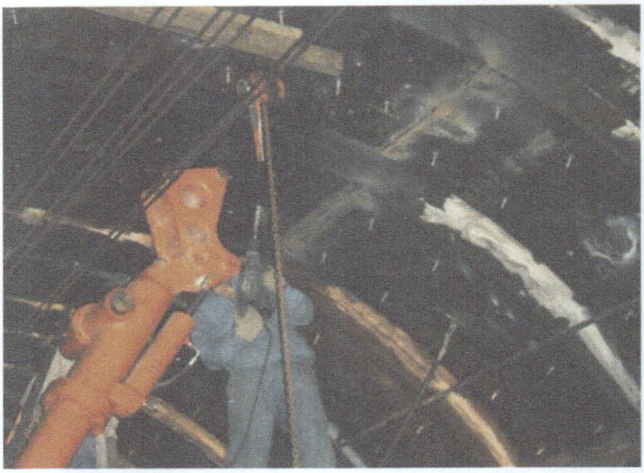

Fig. 5.21 Bolt layout for internal steel ring installation

Fig. 5.22 Completed installation of the internal steel ring

Fig. 5.23 Installation process of steel plates on the track bed

5.2.4.8 Welding of Steel Plate Supports into a Ring

Reliable welding between the segmented steel plates and between the steel plates and corbels is essential to ensure the formation of a robust load-bearing structure that acts in conjunction with the tunnel lining. To meet the requirements of Quality Level 2 according to Chinese Standard for Acceptance of Construction Quality of Steel Structures (GB50205-2020), the welding process includes the following:

(1) The weld joint is prepared with a 55° Y-shaped groove, leaving a 2 mm blunt edge. The groove must be ground, de-rusted, and degreased before welding. Misalignment of the butt steel plates should be avoided during splicing.
(2) CO_2 gas shielded welding is used, with a protective gas mixture of chlorine gas and CO_2. The welding wire diameter is Φ1.2 mm, brand ER50-6, and the welding machine model is Panasonic KR500.
(3) Considering the potential for welding deformation, welding is performed in the following two steps:

    a. After steel plate assembly and electric welding reinforcement, start with the root pass to ensure full penetration of the weld root.
    b. After completing the root pass, weld the joints of one ring of steel plates simultaneously at 2–3 working points. Use multi-layer, multi-pass welding methods, and control the welding current to prevent excessive welding deformation.

(4) To ensure welding quality, the welding location should be well-shielded from wind.
(5) During welding, control the interpass temperature to prevent excessive heating of the steel plates, which could lead to cracking of the tunnel lining on the reverse side of the steel plate.
(6) After welding, conduct ultrasonic flaw detection. Rework any welds that do not meet standards, identify the causes, and make corrections.

5.2.4.9 Welding of Bolts, Handholes, and Dowels

Dowels are installed in the bolt handholes of the tunnel segments within the range of steel plate supports. Typically, $\Phi 12$ mm, $L = 200$ mm threaded steel is used for dowels. Dowel holes are pre-marked on the steel rings, and after the steel plates are in place, welding between the steel plates and the dowels is performed to strengthen the connection between the steel plates and the tunnel segments.

5.2.4.10 Epoxy Resin Filling

After the installation of the steel inner lining, gaps of varying sizes inevitably exist between the lining and the tunnel segments. Filling and bonding these gaps with suitable materials are necessary to unify the segments and the steel lining, allowing them to jointly withstand stress changes.

After installing the steel supports, epoxy resin is injected to fill the space between the circular steel plates and the tunnel segments. Once the steel plates are positioned, the sides of the steel plate arcs are sealed with quick-drying cement at the joints with the tunnel segments. Injection nozzles are installed in the centrally reserved holes in the steel plates (at least two holes per plate, internal thread, 10 mm diameter) and epoxy resin is injected using a small electric grout pump. The injection sequence is from bottom to top, continuing until resin overflows from the upper reserved holes. The process is repeated, and the injection pressure is typically less than 0.1 MPa.

The grouting process is illustrated in Figs. 5.24 and 5.25.

Fig. 5.24 Photo of the epoxy filling construction process

Fig. 5.25 Photo after the epoxy filling construction

5.2.4.11 Non-destructive Testing and Anti-corrosion Treatment

After the installation of the steel hoop is complete, the welds between the segments of the steel lining are subjected to ultrasonic testing to ensure they meet the requirements for Quality Level 2. The interior surface of the steel plates undergoes anti-corrosion treatment with epoxy asphalt coating after the completion of welding and other construction processes. The coating is evenly applied in two layers, ensuring a total thickness of no less than 1.2 mm. As shown in Fig. 5.26, polyurea spray coating is now widely used for corrosion protection. After the steel plates are fabricated in the factory, they are sprayed with polyurea, leaving a 20 cm area around the welding site untreated. This untreated area is sprayed after all welding work is completed to ensure thorough protection.

Fig. 5.26 Photo of steel plate surface painted with epoxy paint

5.2.5 Advanced Nature of the Technique

The technique of reinforcing tunnel linings with internal steel hoops has achieved mechanization, efficiency, and specialized batch construction in tunnel deformation reinforcement. By using structural steel hoops, which work in conjunction with tunnel segments, bolts, and epoxy, this method not only reinforces damaged tunnels but most importantly, enhances the load-bearing capacity and extends the service life of tunnel structures. The steel lining reinforcement process has the following distinct features:

(1) Wide Applicability: Suitable for both under-construction and operational tunnels.
(2) Strong Operability: The construction process is mechanized, efficient, and standardized.
(3) Strong Controllability: Equipment is compact, construction steps are decomposable, and construction time can be precisely controlled.
(4) Significant Effectiveness: Experimental research has shown that after reinforcement, the load-bearing capacity of the entire tunnel structure can be increased by 30–40%.

5.3 Composite Cavity Component Reinforcement Technique

5.3.1 Design Scheme of Composite Cavities

Composite cavity components include the main body of the cavity, composite material layers, fillers, grouting holes, exhaust holes, and bolt holes, as illustrated in Figs. 5.27 and 5.28.

The main body of the cavity can be a metal or non-metal pipe, such as steel, plastic, or composite material pipes. Depending on the shape of the reinforcement surface, it can be composed of one or multiple pipe materials. The surface of the cavity main body is equipped with a composite material layer made of resin and fiber. The interior of the cavity main body is filled with filler materials. The composite cavity components have one or more grouting holes for injecting fillers into the cavity body to enhance the strength of the reinforcement components. The fillers can be materials such as mortar or polymer foams. One or more exhaust holes are also present in the composite cavity components to facilitate the release of air during the injection of fillers.

In practical reinforcement, the composite cavity components are divided into three segments: one upper component and two side components. To reinforce a ring of segments, a total of four rings of composite components plus two at the top are needed, with the rings numbered consecutively as Ring 1, Ring 2, Ring 3, and Ring 4. Each ring is divided into left, middle, and right segments, labeled as GJ-L, GJ-M,

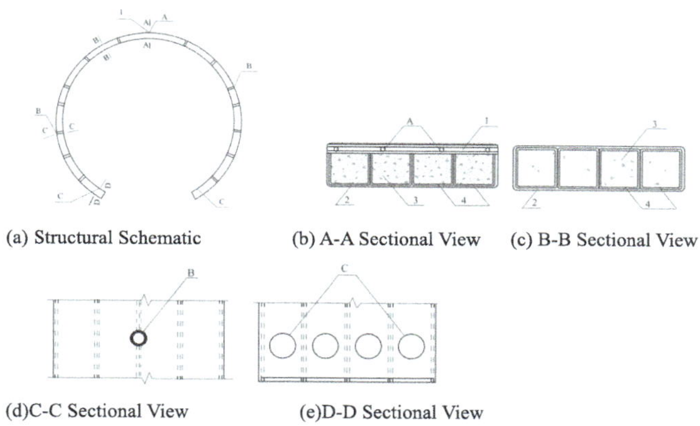

Fig. 5.27 Schematic of typical structure of composite cavity component. *Note* A—exhaust hole; B—bolt hole; C—grouting hole; 1—exhaust passage; 2—composite material layer; 3—filler; 4—cavity main body

Fig. 5.28 Physical model of composite cavity component

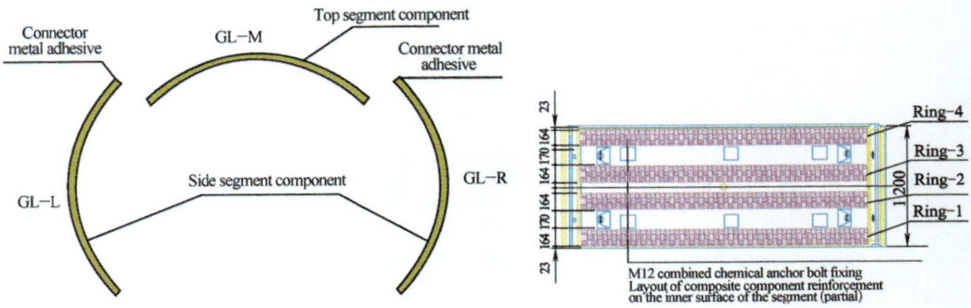

Fig. 5.29 Installation schematic (unit: mm)

and GJ-R, totaling 14 segments. There are also eight pairs of grouting caps and eight sets of bottom jacking devices, as shown in Fig. 5.29.

5.3.2 Construction Process and Key Points of Composite Component Installation

The installation process of composite components is shown in Fig. 5.30.

During the engineering installation and construction, a combined lift truck can be used as a hoist, temporary support, and work platform, as shown in Fig. 5.31.

The installation process is detailed as follows:

(1) Before installation, it is necessary to position according to the benchmark markings, check the fit between the component's arc shape and the tunnel lining, as well as the support and its mounting points. This includes on-site support point cutting, base surface cleaning, component trial assembly, and necessary modifications.

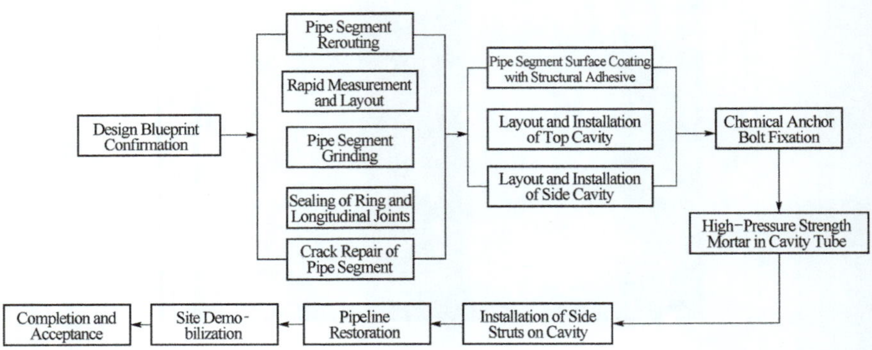

Fig. 5.30 Overall construction process flow

Fig. 5.31 Engineering installation simulation construction

(2) For the installation of the upper segment component, position according to benchmark markings, first apply adhesive on the upper segment component, then lift, temporarily support, and finally fix it in place using a combination of chemical anchor bolts.
(3) During the installation of side segment components, apply adhesive on both the side segment components and at their connection and support points. Then align the connections, position the supports, and form the entire ring. After circumferential pre-tightening, fix it in place using a distribution of combined chemical anchor bolts.
(4) After the installation of the cavity component is complete, fill it with lightweight mortar. Proceed with step-by-step uniform grouting until the grout overflows from the top vent holes. Then close the vent holes with special screws and continue grouting before sealing the grouting ports. The specific operations can be divided into grinding, positioning, pre-assembly, final assembly, and grouting maintenance.

5.3.2.1 Grinding

After the test segment loading is complete, the first step is to grind the internal reinforcement surface. The grinding area should be at least as large as the adhesive area of the composite component. The completion of this process is shown in Fig. 5.32. The specific operations are as follows:

(1) Mark the area on the inner surface of the segment where adhesive is to be applied to define the position.
(2) Use an alternating grinding machine to grind the necessary areas on the inner surface of the segment. The aim is to remove surface laitance and dust, with the grinding depth not exceeding 0.5 mm.
(3) After surface smoothing, use the grinding machine to score and roughen the surface to enhance adhesion. The scoring depth should be about 1 mm, visibly discernible. The spacing between scores should not exceed 100 mm.

Fig. 5.32 Segment marking and grinding

(4) After scoring, vacuum the inner surface of the segment to remove surface dust. Use a vacuum cleaner for this purpose.

5.3.2.2 Positioning and Preliminary Assembly

Using the bottom edge of the tunnel segment as a control boundary, the elevation of each ring of composite components is determined. Each ring is made up of three components, with at least three positioning points for each component. The specific operations are as follows:

(1) Using the bottom edge of the segment as a control point, the lower boundary of the first ring component is 23 mm above the bottom edge; the second ring is 383 mm above; the third ring is 653 mm above; and the fourth ring is 1013 mm above. Mark the positions clearly.
(2) Create extendable bases and extend them to the lower boundary of the first ring's components. Set up three bases for each component, placed at the left, center, and right of the component.
(3) Place the central segment (GJ-M-1) of the first ring on the bases, aligning its central mark with the center of the segment for a trial fit. If the fit is close (gap

less than 1 cm), proceed with the next steps. If the gap is more than 1 cm, try adjusting the component slightly left or right for a better fit.

(4) Fix clamps with a single screw at both ends of the GJ-M-1 component. Insert the left (GJ-L-1) and right (GJ-R-1) components of the first ring into the interface of GJ-M-1. Use a jacking device placed between the processed sleeper and the component, pushing the component tightly against the steel support. Rotate the nuts in the jacking device to adjust the tightness (avoid over-tightening to prevent deformation of the grouting port). Ensure that the bolt holes of the clamps align with those of the left and right components, and that the component fits closely to the segment. The central mark of the central segment should align with the heart center of the top block (deviation less than 1 cm).
(5) Check the fit between the first ring component (Ring 1) and the segment, noting the size of any gaps. Record any gaps exceeding 10 mm and adjust the thickness of the glue in these areas during the subsequent gluing process.
(6) Mark the exact position on GJ-M1 where the component fits tightly and aligns with the heart center of the top block.
(7) After positioning, relax the components onto the platform to ensure they do not fall, then proceed with positioning and preliminary assembly for the second ring.
(8) Positioning and preliminary assembly for the second, third, and fourth rings: The installation positions for these rings have already been determined. Set up temporary bolts at the lower boundaries to support the components. Place five bolt supports under each component of GJ-M, GJ-R, and GJ-L. Add steel plates to the outside of the bolt supports to prevent the components from sliding off. If a ring's components create a large, unadjustable gap with the segment (more than 2 cm), adjust the segment to fit the components. The specific method will be determined on-site based on the size and location of the gap. After adjusting the segment, perform the preliminary assembly for all four rings again.

5.3.2.3 Formal Assembly

For the formal assembly, it's necessary to address any significant gaps identified during the preliminary assembly. The steps are as follows:

(1) For assembling Ring 1, start with the middle segment of the composite component, GJ-M. Evenly spread structural adhesive within the designated gluing area on the segment corresponding to GJ-M (adhesive thickness not less than 4 mm). Then manually apply adhesive on the outer arc surface of GJ-M using plastic gloves, ensuring an adhesive thickness of at least 1 mm (the specifications of the structural adhesive should be the same as those used in the earlier joint tests).
(2) Using supports to hold the component in place, align the centroid mark on the composite component GJ-M during the preliminary assembly with the corresponding heart center at the appropriate height on the top block. Press down firmly, sequentially drill holes in the segment at the pre-marked bolt holes on the component, clean the holes, and insert combination chemical anchors. Tighten

the composite component immediately with an electric wrench. The order of operation for the anchor bolt holes is MS-14, MS-15, MS-13, MS-16, MS-12, MS-17, MS-11, MS-18, MS-10, and MS-19, as shown in Fig. 5.33.

(3) Remove the first ring's left component GJ-L-1 and apply adhesive.
(4) Apply adhesive to the corresponding pipe piece area of the left component GJ-L-1 of the first ring (adhesive thickness > 4 mm), and apply adhesive to both inner and outer contact surfaces of the outer arc and pipe spigot joint of the component (adhesive thickness > 1 mm).
(5) Fix the clamp to the left end of the middle component GJ-M-1 with a single screw.
(6) At the appropriate installation height, slightly tilt the left component GJ-L-1, first insert its joint into the clamp, and then press GJ-L-1 firmly onto the pipe piece structure.
(7) Install the bottom jacking device for the left section component GJ-L-1 (device height: 85–120 mm, adjustable up to 35 mm), leaning against the track bed (if not leaning, add multiple steel plates or tiles to ensure it leans), then adjust the jacking device to press against the left section component GJ-L-1, ensuring the joint of the left component is tightly pressed against the middle component (do not overtighten to prevent deformation of the grouting port). Check if the component aligns with the pipe piece. Verify if the bolt holes in the joint clamp align with the left component. If aligned, first fix the corresponding reserved anchor bolt holes in the designated order, starting from near the middle component and moving outward, i.e., first MS-9, MS-8, MS-7, MS-6, MS-5, MS-4, MS-3, MS-2, and MS-1. After fixing the chemical combination anchor bolts, remove the clamp, and apply 2 mm thick adhesive on the inside and 3 mm thick adhesive on the outside of the clamp. Screw the clamp back onto the joint of the composite component.
(8) Install the bottom jacking device for the right section component GJ-R-1 (device height: 85–120 mm, adjustable up to 35 mm), leaning against the track bed (if not leaning, add multiple steel plates or tiles to ensure it leans), then adjust the jacking device to press against the right section component GJ-L-1, ensuring the joint of the right component is tightly pressed against the middle component (do not overtighten to prevent deformation of the grouting port). Check if the component aligns with the pipe piece. Verify if the bolt holes in the joint clamp align with the right component. If aligned, first fix the corresponding reserved anchor bolt holes in the designated order, starting from near the middle component and moving outward, i.e., MS-20, MS-21, MS-22, MS-23, MS-24, MS-25, MS-26, MS-27, and MS-28. After fixing the chemical combination anchor bolts,

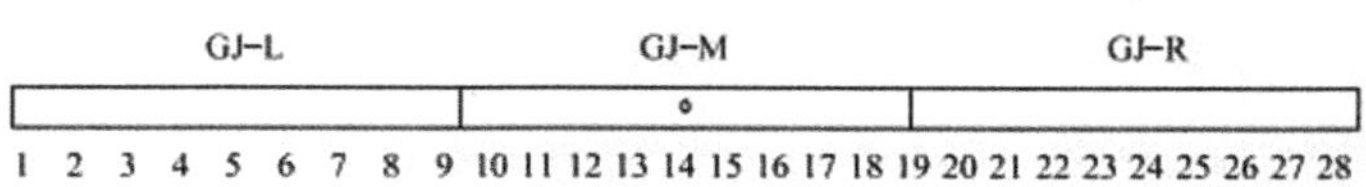

Fig. 5.33 Schematic diagram of chemical anchor bolt holes

remove the clamp and apply 2 mm thick adhesive on the inside and 3 mm thick adhesive on the outside of the clamp. Screw the clamp back onto the joint of the composite component. After completing the assembly of Ring 1, proceed upward to complete Ring 2, Ring 3, and Ring 4. The construction process is shown in Fig. 5.34.

5.3.2.4 Grouting of Composite Components

After installation, grouting work for Ring 1 is carried out first. The grouting material is C60 high-performance steel pipe grout, and unidirectional piping measures should be used. The grouting is performed from the lower right end to the left end, with the grouting pressure not exceeding 1.5 MPa (not more than 1.0 MPa for horizontal installation). When grout emerges from the small cavity of the steel pipe, the corresponding grout outlet valve is closed until all four grout outlets are closed. Then, continue to maintain grouting until the pressure reaches 2 MPa, at which point the grouting valve is closed. The amount of grout is controlled by the flow meter, and a site photo is shown in Fig. 5.35. The same operation is followed to complete the grouting work for Rings 2, 3, and 4 in sequence. The grouting equipment is placed outside the segment, and the grouting pipeline is no less than 5 m in length. The steps are as follows:

(1) Manual mixing of grout at the site.
(2) Pouring into the grouting machine using a measuring bucket.
(3) Grouting unidirectionally from the right end to the left end of the first ring.
(4) Strictly controlling the pressure and flow rate during grouting.
(5) When grout starts spraying from the right end exhaust hole, judge the amount of grout and compare it with the volume of the component's internal cavity. Close the corresponding grout outlet valve until all four grout outlets are closed, then continue to maintain grouting until the pressure reaches 1 MPa, then close the grouting valve, completing the grouting.

Grout the components of Rings 2, 3, and 4 in the same manner in sequence, ensuring that the grouting volume is close to the volume of the component's cavity.

5.3.3 Advancement of the Technique

The reinforcement construction using composite cavity components is space-efficient and allows for quick installation at multiple points and surfaces, significantly enhancing construction efficiency and practicality. The advanced features are as follows:

(1) Using on-site measurements to customize components in the factory ensures an effective fit between the original structure and the reinforcement components. The use of special high-strength adhesives and pre-expanded combination

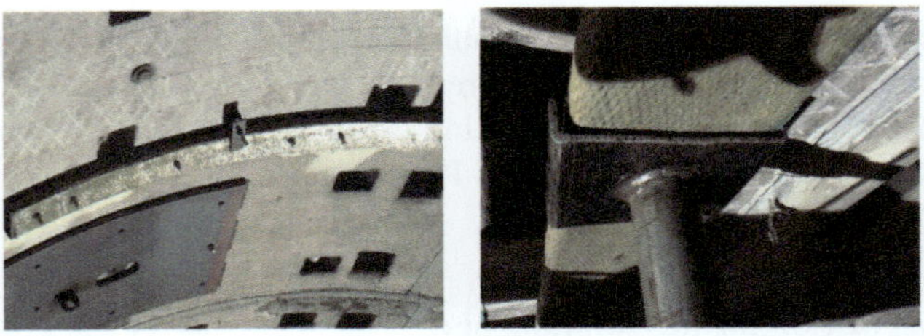

(a) Support Bracket (b) Top Pressure Plate

(c) Applying Structural Adhesive (d) Sealing Edges with Structural Adhesive

(e) Composite Component Installation (f) Drilling Holes

(g) Injecting Chemical Adhesive (h) Top Jacking Device

Fig. 5.34 Construction site photos

(a)Grouting Ball Valve (b)Pipeline Connection

Fig. 5.35 Grouting construction site photo

chemical anchor bolts with immediate pressure-bonding techniques allows the new and old structures to work together, effectively suppressing further tunnel deformation. This "combined effect" significantly enhances the load-bearing capacity of the original structure.

(2) Due to the narrow space in subway tunnels, especially in operational subway tunnels, where the construction window time is extremely limited each day, the reinforcement construction using composite cavity components does not occupy space and can be quickly installed at multiple points and surfaces, greatly enhancing the constructability.

(3) Composite cavity components effectively reduce the susceptibility of concrete segments to deformation due to changes in the tunnel load environment and the variable characteristics of the multi-segmental ring structure.

Chapter 6
Double-Liquid Micro-disturbed Grouting Technology for Deformation Control

6.1 Introduction

Shortly after the Shanghai Metro Line 1 commenced operation, noticeable differential settlement occurred in some sections of the tunnel. In some cases, it exceeded the allowable adjustment range for track fasteners, necessitating the re-customization of fasteners to meet clearance requirements. The longitudinal differential tunnel settlement unavoidably caused the joint dislocation of tunnel segment rings, with worsening joint dislocation leading to the failure of sealing gaskets, aggravated water leakage, disengagement between the track bed and tunnel segments, and in turn faster differential settlement. This, consequently, posed a threat to the tunnel's structural integrity and train operation safety.

In addition, various reasons also sparked convergence and deformation in some tunnel sections. Excessive transverse convergence and deformation can lead to the expansion or compression of tunnel segments, resulting in issues such as water leakage, segment compression damage, bolt failure, and separation of the track bed from the segments, thereby affecting structural integrity and train operation safety. When the cumulative value of tunnel deformation exceeds a certain threshold, it can result in severe tunnel damage and even a total collapse of the tunnel.

Both of these major deformation diseases can be solved by reinforcing the ground around the tunnel. In the late 1990s, grouting was conducted to adjust the tunnel alignment, but no effective solution was found for differential settlement with large curvature, great amplitude, and high settlement rate. The settlement of tunnels in saturated soft soil still cannot be controlled effectively, and therefore, the traditional compensation grouting, such as split or compaction grouting, is commonly used in China and other countries. But traditional grouting runs into three common problems:

R. Wang et al., *Structural Safety Early Warning and Treatment Technology of Metro Tunnel in Soft Soils*, Key Technologies for Tunnel Construction under Complex Geological and Environmental Conditions,
https://doi.org/10.1007/978-981-96-6410-8_6

(1) Massive disturbance. There may be substantial heave during the grouting process, followed by even greater settlement.
(2) Poor controllability. It is hard to control the flow path in the soil during the grouting process.
(3) Difficulties in guaranteeing the effectiveness of grouting. It is challenging to quantitatively or even qualitatively predict the grouting effectiveness.

Furthermore, traditional grouting methods are short of industry construction specifications, theoretical guidance, and normative operational instructions. In this context, there is a deficiency in successful grouting projects and engineering experience. Moreover, implementing grouting within operational tunnels faces numerous practical challenges and constraints, such as restricted tunnel equipment clearances, limited workspace, tight schedules due to the short window time, and induced additional structural internal force demands. All traditional grouting methods are inapplicable and can potentially have adverse effects on the tunnels in such conditions.

After years of testing and study, Shanghai Metro, focusing on the "poor results, significant disturbance, and poor controllability" in grouting, has pioneered a reliable and effective grouting method for controlling settlement, and for convergence and deformation in soft soil tunnels, called the micro-disturbed grouting technology.

6.2 Technology Development Process

6.2.1 Research and Development Approach

6.2.1.1 Research Objectives

The target is to develop a micro-disturbed grouting process that can reinforce the underlying stratum of the operating tunnels within a limited range without impeding the operation of trains, so as to control the further development of tunnel settlement with higher effectiveness, better controllability, and lower disturbance.

6.2.1.2 Investigation of Grouting Process

Before the on-site tests, long-term research and analysis were conducted on existing grouting methods, equipment, and processes to clarify the gap between traditional grouting and high-standard requirements.

6.2.1.3 Confirmation of the Testing Scheme

Various experiments need to be done to confirm the key technologies and equipment required in the grouting process. Through experiments, six aspects of research need to be completed: research and development of supporting equipment, date analysis of construction guide, and confirmation of grouting material parameters, of grouting process parameters, of grouting disturbance effect, and of soil reinforcement. The specific technical roadmap is shown in Fig. 6.1.

6.2.2 Testing Process

Since 2008, in-depth analysis has been conducted on existing grouting methods, processes, and equipment, especially the grouting effects and disturbances. The aim is to identify the gap between traditional grouting and the requirements for tunnel settlement control, determine the research directions and objectives, and formulate and improve research plans. Research is conducted in confirming engineering testing sites, testing content, equipment development, testing parameters, testing processes, on-site testing, data analysis, and process parameters improvement. The experimental

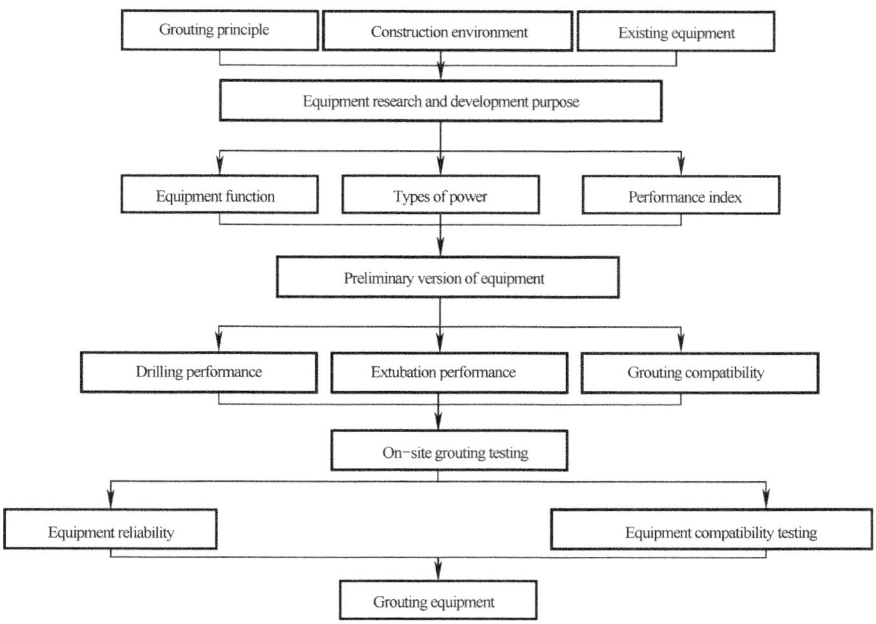

Fig. 6.1 Technology roadmap

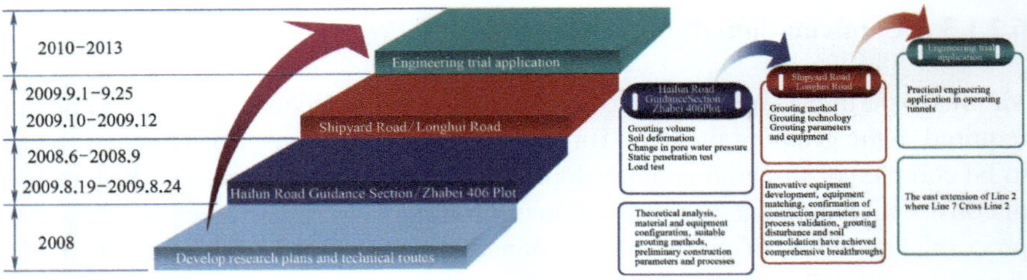

Fig. 6.2 The experimental process and the main content of each stage

process and the main content of each stage are shown in Fig. 6.2. To achieve desired experimental effects, on-site tests are carried out in three stages.

6.2.2.1 Phase 1 of Preliminary Testing

Since June 2008, on the basis of theoretical and technical discussions, technical approaches of testing and design research have been decided. The first in situ tests were conducted in the section of "Hailun Road Station-Baoshan Road Station" along Line 4 (Jun–Sep 2008) and Lot 406# near the tunnel of Hanzhong Road of Line 1 during Aug 19–24, 2009, as shown in Fig. 6.2. The main purpose of this stage is to understand the increased strength of the soil after grouting, confirm the effective reinforcement range, and determine appropriate grouting methods, processes, parameters, and equipment. The test contents include: grouting volume test, grouting stages test, cone penetration test (CPT), and load test. The analyses of soil deformation and changes in pore water pressure after can optimize construction parameters; CPT can determine the grouting diffusion range; Load tests can verify the grouting effect. The research in this stage enables us to access basic information, such as grouting materials and equipment configuration, grouting methods, processes, and construction parameters. Grouting equipment confirmed after the preliminary testing is shown in Fig. 6.3.

6.2.2.2 Phase 2 of In Situ Testing

Since July 2009, further experimental research has been conducted on equipment innovation, equipment matching, construction parameters, process testing, disturbance effect, and related aspects. Testing locations were determined at Chuanchang Road from Sep 1 to 25, 2009 and Longhui Road (from October 2009 to December 2009), as depicted in Fig. 6.4. The aim of this phase was to identify suitable grouting methods, processes, parameters, and equipment, providing experience for future grouting reinforcement within operational tunnels. Additionally, various measurements were employed to monitor the disturbance of surrounding soil and subsequent settlement impact caused by grouting, confirming the increase in soil strength after

(a)Injection pipe (b)Grouting sprinkler (c)Spray test

(d)Drill pipe going down (e)Slurry stirring (f)Grouting pressure gauge

(g)Grouting test

Fig. 6.3 Grouting test in Phase I

grouting and the effective reinforcement range. Through the experiments in this phase, comprehensive breakthroughs were achieved in design, processes, parameters, equipment, technologies, and devices. These achievements underpinned the formal grouting within operational tunnels. The improved equipment is illustrated in Fig. 6.5, and on-site sampling of the reinforced soil in Fig. 6.6.

6.2.2.3 Phase 3 of Application and Continuous Refinement of Micro-disturbed Grouting

Grounded on the previous extensive in situ tests, settlement grouting control tests were conducted within tunnels. From May 2010 to May 2011, the tunnels of Line 2 achieved excellent control effect, marking the initial success of the micro-disturbed grouting technology. Subsequently, grouting lifting was applied in various sections, including transit areas between Line 7 and Line 2, between Line 11 and Line 2,

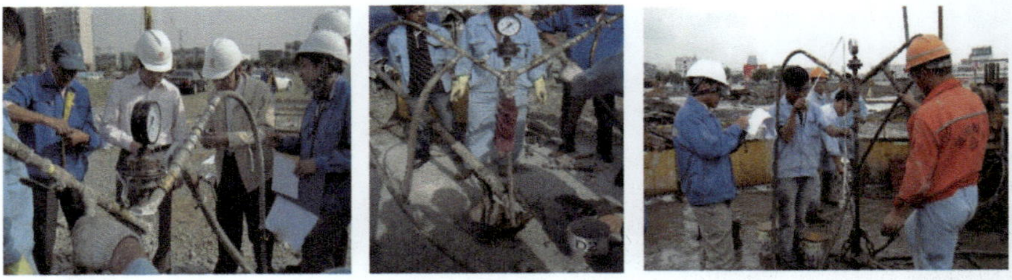

Fig. 6.4 Second stage on-site testing

Fig. 6.5 Improved grouting equipment

Fig. 6.6 Soil after grouting reinforcement

between Line 11 and Line 1, eastern extension of Line 2, Hailun Road guiding section, interconnection passage of Line 9, and others. By now, the micro-disturbed grouting technology is continuously refined in various aspects such as equipment, process technology, and construction parameters.

The technology was first introduced to address differential settlement in tunnels. Then, to perform grouting treatment, it was elaborately designed and refined before being applied to the Xinjiangwancheng in Line 10 in November 2009 and later to Dong'an Road in Line 7, Hainan Road in Line 10, Sanxiang Project in Line 2, and Gucun Park in Line 7, achieving outstanding control effect.

6.3 Technical Principles

6.3.1 Treatment of Longitudinal Differential Settlement

The direct cause of longitudinal differential settlement is that the underlying strata of the tunnel are relatively weak and have low carrying capacity, undergoing differential deformation and significant displacement under the influence of the non-uniform load. To control tunnel settlement, efforts must be made to improve the physical and mechanical properties of the underlying strata and enhance their strength (Yan and Wang 2018).

Since tunnels are assembly structures with numerous joints, there is a reduction in stiffness at these joints, and a slight joint dislocation is allowed between rings. The longitudinal differential settlement in tunnels primarily caused dislocation of the circumferential joints. To control the settlement, given the allowable slight joint dislocation between circumferential joints, micro-disturbed grouting is applied to restore the segments without any damage to the tunnel structure and to bend the alignment through sustained small displacement superposition. Each grouting operation causes insignificant disturbance to tunnel lining and surrounding soils, and cumulative grouting is helpful to restore the displacement of a tunnel section.

The specific method for micro-disturbed grouting is as follows: based on the monitoring of tunnel settlement and the empirical formula of settlement in Shanghai soft strata, it is possible to predict the future settlement of tunnels and divide the grouting scope into zones and stages to implement grouting. During grouting, grouting holes are arranged at certain intervals longitudinally along the tunnel sections, and each hole is layered and overlaid with grout from the bottom of tunnels in stages, small amounts, and by multiple times. It is important to control the amount of grout and reduce the disturbance to the strata, ensuring that the tunnel uplift driven by each grouting of ΔS_1 is greater than or equal to ΔS_2, which is the sum of tunnel settlement due to the consolidation of the foundation during the intervals between two grouting operations, as shown in Figs. 6.7 and 6.8. The consolidation settlement is caused by disturbance of soil structure and partial dissipation of excess pore pressure.

The grouting process for each grouting hole is roughly divided into three stages: uplift grouting, interval grouting, and consolidation stable grouting.

In general, during the uplift grouting stage, tunnel uplift in the control section must reach certain expected value. To achieve this, during this stage, the interval time Δt for each grouting operation is reduced to increase the effective uplift amount of each grouting operation ($\Delta S_1 - \Delta S_2$). The total tunnel uplift caused by each grouting operation in this stage should reach the desired value, denoted as

$$R = \sum(\Delta S_1 - \Delta S_2). \tag{6.3.1}$$

When the expected value is reached, grouting is paused for a period of time to release the excess pore pressure triggered by grouting in the underlying strata, so that

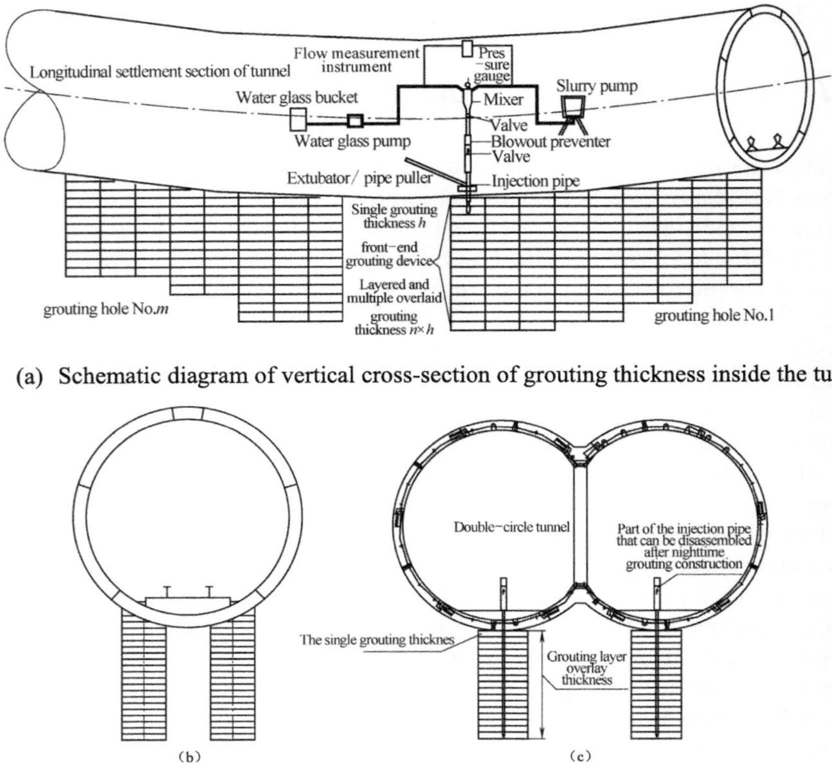

Fig. 6.7 Grouting diagram

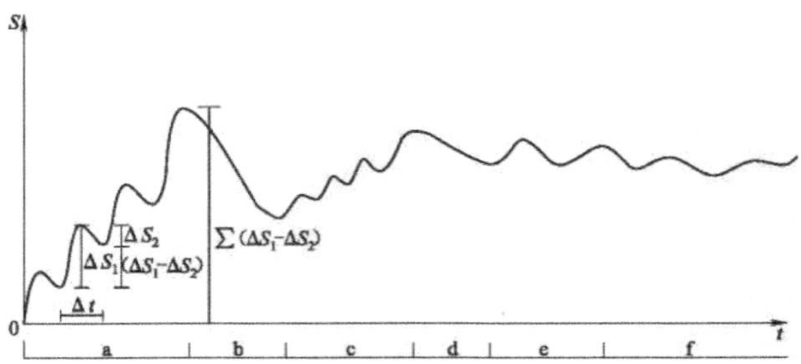

a—the initial stage of the uplift grouting; b—stop grouting; c—the second stage of the uplift grouting; d—stop grouting; e—the third stage of the intermittent grouting;f—the secondary consolidation stage of the intermittent grouting

Fig. 6.8 Schematic diagram of the uplift and followed subsidence of staged grouting

the tunnel foundation undergoes certain consolidation settlement. Experimental tests have shown that the complete dissipation of pore pressure in mucky soil and clayey layers takes about 1 week. If it is necessary to further increase the total tunnel uplift, the second stage of interval grouting begins. When the grouting uplift reaches the expected value, the second stage will stop. After another consolidation settlement of the strata, the rate of tunnel settlement begins to decrease. If the settlement rate exceeds 0.02 mm/d, the third stage of consolidation stable grouting is initiated to maintain relatively stable tunnel settlement. During this stage, the interval between each grouting operation can be appropriately extended, but the grouting depth remains in the strata below the tunnel that require reinforcement. The consolidation stable grouting stops until the tunnel settlement rate is below 0.01 mm/d.

In each grouting stage, the interval time for each grouting operation varies depending on the geological conditions. In mucky soil and clayey layers, Δt in the initial stage is 2–3 days, in the second stage 5–7 days, in the third stage 2–3 weeks, and in the second consolidation settlement stage can be extended to more than 1 month.

6.3.2 Control of Large Traverse Convergence

Normally, tunnels remain stable under the loads from the bottom, top, and lateral soil. However, when there is an increase in vertical pressure such as overburden pressure on the tunnel roof or a decrease in horizontal load such as the evacuation and unloading of the foundation pit on one side of the tunnel, the tunnel convergence will occur with horizontal diameter of the tunnel lining rings increase. In terms of overall tunnel stability, lateral unloading is more sensitive than upper loading, as confirmed by full-scale ring loading and unloading tests. The magnitude of convergence and deformation mainly depends on the changes in the horizontal stress coefficient (λ) and the lateral resistance coefficient (P_k), that is, the decrease of λ and the loss of P_k. λ is generally around 0.65–0.75, P_k ($P_\mathrm{k} = k\delta$) relies on the severity of disturbance. For example, k will plummet to 0 under severe disturbance. With λ below 1.0, objectively, the increased loads on the top of tunnels will hamper the tunnel convergence. Therefore, given the control over convergence of tunnels, it is necessary to first decrease a portion of the loads on top of tunnels while increasing lateral pressure on the tunnel, thereby improving the values of λ and P_k.

To maintain a perfect circularity of the tunnel lining ring, it is necessary to adjust λ and P_k. Therefore, the general idea of micro-disturbed grouting control for convergence is as follows. Implement single-row or multi-row double-fluid grouting on one side or both sides of the tunnel. It can rapidly generate excess pore pressure to reduce the transverse convergence of tunnels, and meanwhile, the segments will experience minor disturbances, including translation, rotation, and deformation, due to the well control of grouting pressure and pore pressure,. As time progresses, the dissipation of excess pore pressure caused by grouting and the reconsolidation of the surrounding strata will lead to an increase in transverse convergence of the tunnel. By appropriately utilizing the squeezing effect caused by grouting and the deformation during the dissipation of excess pore water pressure, and by effectively arranging the grouting sequence around the tunnel, the reduction in transverse deformation during grouting should be greater than the increase in transverse deformation caused by soil consolidation and the subsequent dissipation of pore pressure. Ultimately, this approach aims to reduce the tunnel convergence through grouting (Zhang et al. 2014).

Controllable grouting disturbances, when implemented effectively and repeatedly, can significantly improve the surrounding soil conditions of the tunnel, enhance lateral resistance, and reduce transverse convergence of the tunnel under grouting pressure and consolidation, therefore getting desired control effect. Taking advantage of the reversible deformability of lining rings assembled through straight joints, orderly and controllable grouting is conducted to the corresponding layers on both sides of the tunnel lining. The grouting is carried out in an "even, small-amount, multiple-point, and multiple-time" fashion to achieve orderly the recovery of tunnel convergence. The recovery of tunnel convergence can improve the perfect circularity of the tunnel lining ring and adjust the stress state of the tunnel structure and thus enhancing the overall safety of the tunnel lining ring structure.

When large traverse convergence occurs outside the tunnel, given the on-site construction conditions, grouting holes will be arranged on one side or both sides, typically in 2–4 rows. The total grouting volume is determined by the actual grouting effectiveness. The fundamental principle is consistent with that in differential settlement control. The lateral grouting will increase the stiffness of the lateral soil and enhance lateral horizontal resistance. Supported by the following injection of cement slurry in certain pressure, the lateral tunnel segments will be squeezed to experience inward convergence and deformation, reducing large convergence and joint openings, controlling the development of convergence, and improving the internal force conditions of the tunnel segments to some degree.

The arrangement of grouting holes and the grouting scope is as shown in Figs. 6.9, 6.10 and 6.11.

The following experience can be referred with the basis on the lateral grouting practice for Shanghai subway tunnel.

(1) Grouting holes are arranged horizontally in 2–4 rows on one side or both sides, parallel to the tunnel axis. The distance of each row is no less than 3 m, and that of each grouting hole is 1.2 m.

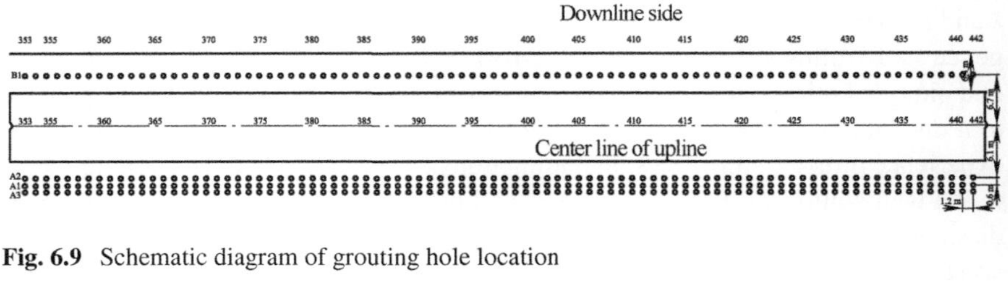

Fig. 6.9 Schematic diagram of grouting hole location

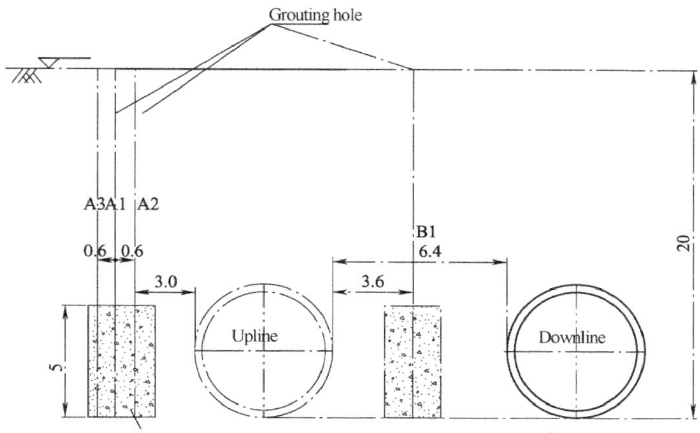

Fig. 6.10 Section of tunnel grouting hole location (unit: m)

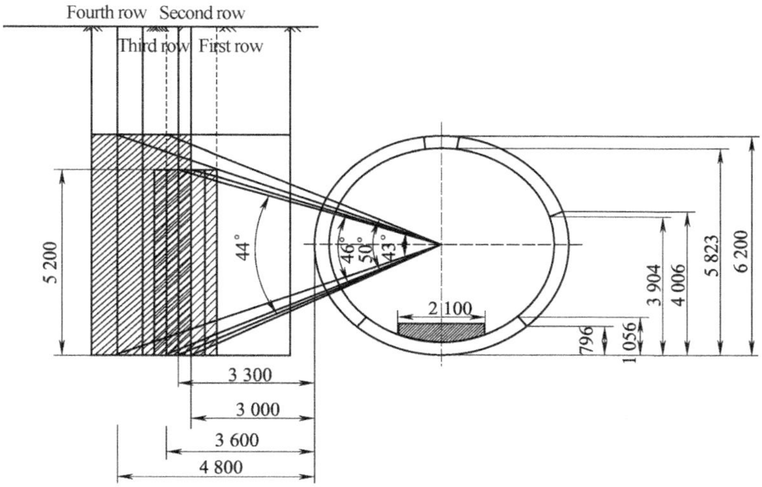

Fig. 6.11 Schematic diagram of pile section within the grouting range (unit: mm)

(2) The grouting depth for each hole corresponds to the soil height from the top of the tunnel to the bottom of the tunnel, determined by the soil conditions and tunnel convergence.

When grouting is performed on both sides of the lining rings, the segments will experience inward movement and deformation influenced by excess pore pressure, while the sealing block at the crown will move outward and deform. This results in an overall reduction in the convergence of the lining rings and reasonable internal force distribution of tunnel segment. The excess pore pressure will evidently exist for about 2–5 days. As the pressure dissipates, there is a trend of the tunnel returning to its previous state in terms of tunnel convergence. Through experiments, it is possible to accurately determine the time interval between two grouting operations, allowing for a certain amount of convergence recovery. The strategic implementation of longitudinal jump grouting along the tunnel without excessive overlap can reduce transverse convergence gradually and controllably, restoring it to the design conditions.

6.4 Grouting Techniques

6.4.1 Grouting Materials and Parameters

In uplift grouting for restoring tunnel settlement, since the uplift ΔS_1 and settlement ΔS_2 largely depends on the material properties injected into the soil, micro-disturbed grouting requires that the grout injected into the soil generates uplift ΔS_1 in fair proportion while keeping ΔS_2 as small as possible. It is essential that the grout injected into the soil exhibits high density, rapid setting, small volume shrinkage, and high strength. Therefore, given the excellent rapid setting and high strength, it is necessary to select cement-silicate double solution grouting as grouting material. To achieve smooth injection pumping and increase ΔS_1, numerous experiments have showed that the water–cement ratio of the cement slurry should range from 0.6 to 0.7, the ratio of cement slurry to water glass from 2.5:1 to 3:1.

In each grouting process for each hole, as the core technique of the process, strict control of grouting parameters decides the final result of grouting. The technical parameters are determined by the geological conditions and tunnel settlement characteristics. Based on multiple experiments and engineering practice results, the technical parameters used in Shanghai for flowing mucky plastic clay are shown in Tables 6.1 and 6.2.

6.4.2 Grouting Equipment

The equipment used in the micro-disturbed grouting process is shown in Fig. 6.12, including:

(1) Injection pumps, including cement slurry pumps and water glass pumps, both of which need to simultaneously have good grouting controllability, small and stable flow rates, adjustability, continuous and even slurry delivery, small and lightweight in size, suitable for handling and construction in narrow spaces, etc.
(2) Mixer: It can evenly mix the two types of grouting slurries, one end connected to the cement slurry and water glass delivery pipe, and the other end connected to the injection pipe. An external pressure gauge is attached to the mixer to promptly reflect the double-liquid slurry pressure and ensure that there is no

Table 6.1 Longitudinal differential settlement grouting parameters

Item	Specific parameters
Grouting material	• Water–cement ratio for cement slurry: 0.6–0.7 • Water glass density: 35–42° • Water glass modulus: 2.85–3.2 • Ratio of cement slurry to water glass: ideally controlled between 2.5:1 and 3:1 • Dual-fluid grout initial setting time: approximately 20 s
Single-time grouting length	• The length of a single-time grouting should not be excessively long • Typically, the length of a single-time grouting is between 88 and 352 mm
Single-time grouting volume	The single-time grouting volume should be determined by the grouting length and admixture. It should not exceed 80 L
Grouting flow rate	Generally, the double-liquid slurry flow rate is 20 L/min, with a cement slurry pump flow rate of 14–16 L/min and a water glass pump flow rate of 4–6 L/min
Pulling speed	Determined by the single-time grouting volume, grouting length, and double-liquid slurry flow rate
Borehole spacing requirement	Spaced borehole construction is used in longitudinal tunneling, with intervals not less than the length of 2 ring segments
Time interval of grouting cycles	In mucky soil soft clay, the time interval generally takes 1–3 days for the uplifting grouting phase, 2–3 weeks for the interval grouting phase, and more than 1 month for the consolidation and stabilization grouting phase

Table 6.2 Parameters for large traverse convergence of lateral grouting

Item	Specific parameters
Grouting material	The water–cement ratio is 0.6–0.7, the modulus of water glass 35–42, and the modulus 2.85–3.2. The ratio of cement slurry to water glass should be controlled between 2:1 and 3:1, and the initial setting time for double-liquid slurry is approximately 20 s
Single hole grouting volume	According to a 20% mixing ratio, the grouting volume per meter length is approximately 226 L, with a grouting time of 10 min. The flow rate for cement slurry is 15 L/min, and for water glass, it is 7.5 L/min, totaling 22.5 L/min
Grouting flow rate	Typically, the flow rate for double-liquid slurry is 20 L/min, with the cement slurry pump flow rate at 14–16 L/min and the water glass pump flow rate at 4–6 L/min
Pulling speed of grouting pipe	The pipe should be pulled up 50 mm every 30 s
Borehole spacing requirement	In the same row, construction should follow a “skip one every two” pattern, with no less than a two-day interval between adjacent hole grouting

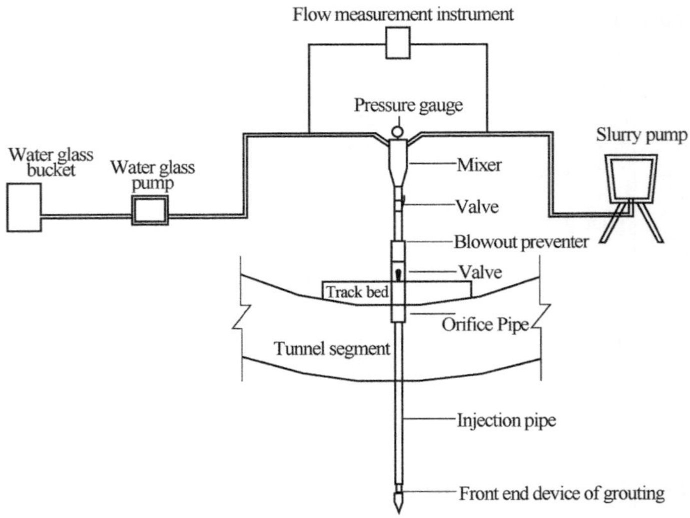

Fig. 6.12 Schematic diagram of the micro-disturbed grouting implementation device

residual coagulated slurry inside the mixer and various pipelines, thus guaranteeing smooth and even inflow and outflow of the slurry. The initial coagulation time of the double-liquid slurry is usually 16–24 s, and the mixer's function is to thoroughly mix the unsolidified slurry before entering the strata. In this way, the slurry will quickly achieve initial coagulation after entering the strata, thereby realizing the proper lifting of the tunnel and perturbation to the soil.

(3) Flow meters: They are connected to the slurry delivery pipeline to record and monitor the real-time flow rate of the slurry, calculate its cumulative volume, and accurately control the slurry volume for each injection into the strata.

(4) Extubation devices: Fig. 6.13 presents the structure of a simple extubation device, which is easy to carry in operating tunnels. Assisted by a measure tape, operators can pull the pipe evenly.

(5) Front-end grouting devices, as shown in Fig. 6.14: The front-end grouting devices are equipped with two rows of plum blossom-shaped grouting holes, each with a diameter of 4 mm, evenly distributed in each row. Before grouting, protective covers are used to protect the grouting holes to prevent them from being clogged by soil during the descent of the pipe. During grouting, the protective covers are removed, and all eight grouting holes are opened simultaneously to achieve even injection of the slurry.

In recent years, through the summary of grouting engineering cases, grouting equipment has been continuously improved, and manual operation is shifted to equipment integration. For example, separate drilling and hole-forming systems are updated into integrated "drill-inject-pull" grouting systems, as shown in Fig. 6.15. Simultaneously, active and rapid intervention grouting system components have been

Fig. 6.13 Extubation devices for lateral grouting

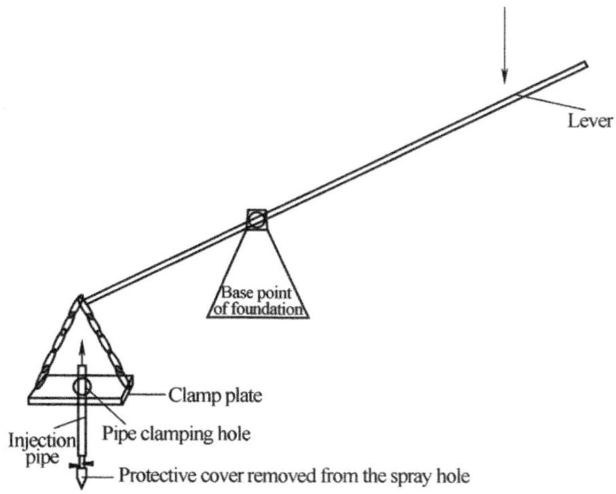

Fig. 6.14 Schematic diagram of the front-end device

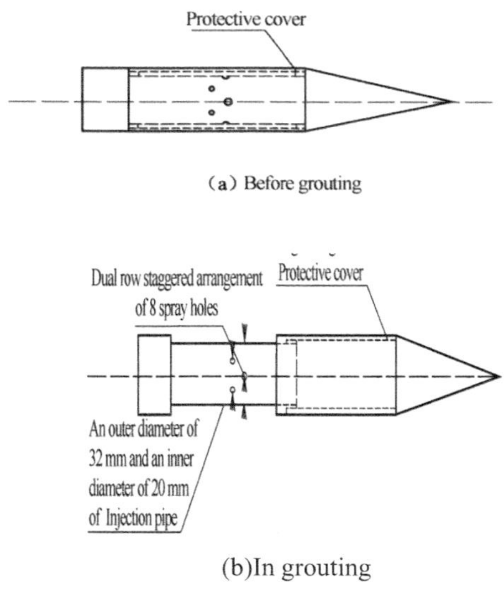

developed to adapt to various complex construction conditions and improve work efficiency. These components include a tracked mobile chassis, which can not only adapt to various complex terrains but also move rapidly; full hydraulic control and instrument display for convenient and flexible operation; and hard alloy ball-tooth drill bits that can easily penetrate hard shell layers.

(a) Tracked mobile chassis (b) Full hydraulic control and instrument display (c) Hard alloy ball-tooth drill bits

Fig. 6.15 Integrated "drill-inject-pull" grouting system

6.4.3 Technological Process

6.4.3.1 Vertical Grouting for Differential Settlement

During the first single-time grouting in holes, when the injection pipe connected to the front-end protection device is pressed into the strata at a depth of $H = h + C$, an extubation device is used to pull the pipe up. Once the pipe is pulled to the height of C, the front-end protection device automatically disengages to ensure that all grouting holes are opened at the same time. Grouting is initiated at the preset flow rate, and as the pipe is pulled upward, grouting continues. When the pipe is raised to a height of $h + C$, grouting is halted, and at this point, the grouting height is h, completing the first-time grouting operation. After the slurry pressure dissipates and the slurry is condensed (about 15 min), the grouting front-end device is pulled out along with the injection pipe. During the second-time grouting, the injection pipe is pressed into the depth of $H = h + C$, repeat the process of the first-time grouting, and overlap the grout in two times along the vertical direction. This procedure is repeated for the third-time, fourth-time, and subsequent nth-time grouting operations, with strict control over the grouting volume and evenness for each operation to minimize disturbance to the surrounding soil. The time intervals between each grouting operation are determined by the parameters set for each grouting stage, as shown in Fig. 6.16.

The grouting in each hole is carried out in stages and in sections according to its location and tunnel settlement prediction. Predictions are made for future settlement development of the tunnel, according to the settlement monitoring and rates, and empirical settlement formulas. Based on the tunnel settlement prediction curve

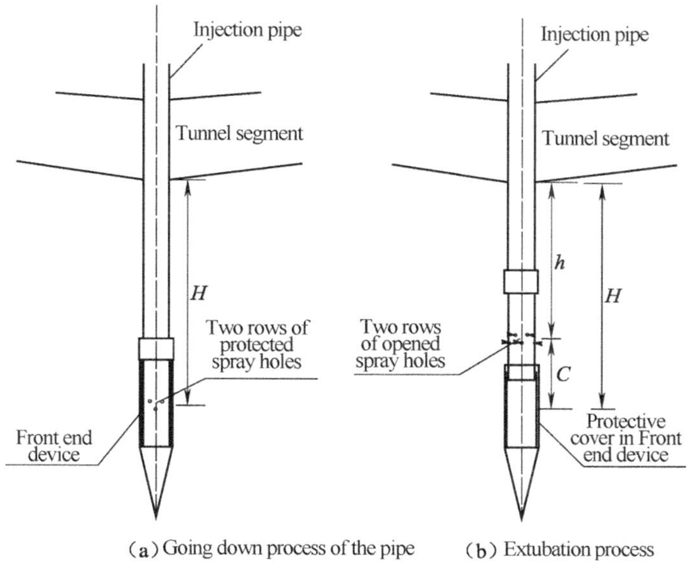

Fig. 6.16 Schematic diagram of single grouting

and the degree of differential settlement, it is impossible to determine the grouting reinforcement scope and control objectives, define different grouting segments, and decide phased grouting plans for each grouting hole within these segments along with the corresponding grouting intervals.

6.4.3.2 Lateral Grouting Process for Large Transverse Convergence

This process primarily includes grouting holes layout, drilling and installation of injection pipes, grout mixing and pipe connection, grouting, and completion of grouting.

(1) Setting out grouting holes: Based on the design, the positions of grout holes are surveyed and marked by monitoring teams on-site, and the ground elevation and grout depth for each grout hole are determined.
(2) Drilling and installation of injection pumps: Atlas equipment is used to drill grout holes that penetrate through hard ground surfaces and building backfill layers. The borehole depth is generally 7–10 m, allowing for the installation of injection pipes. The grouting core pipe adopts a ϕ 28 mm seamless steel pipe connected by a threaded mouth. Several small holes are drilled on the side of the core pipe, with hole diameters ranging from 1 to 4 mm, and protective hole devices are installed to prevent clogging.

(3) Connection of injection pipelines: The injection pump, flow meter, mixer, and injection pipes are connected through the injection pipelines, and a backflow check ensures that the pipelines are unobstructed.
(4) Slurry preparation: Cement slurry is prepared using a mixing system according to the water–cement ratio parameters.
(5) Grouting and pipe pulling: The micro-disturbed grouting is carried out using a double pump and double-liquid grouting method. A specialized pipe-pulling device is used to simultaneously grout and pull the pipe. This process is performed slowly, continuously, and evenly. The pipe-pulling speed matches the grouting flow rate, grouting single-section height, and grout volume, approximately pulling 50 mm every 30 s.
(6) Pulling out of injection pumps: After completing the grouting as required, the injection pump is left to stand for approximately 10 min. Once the grout starts to set, a dedicated pipe-pulling device is used to pull out all the injection pipes, completing a single grouting operation.

6.5 Technological Advancements and Applicability

The micro-disturbed grouting reinforcement technology represents a groundbreaking advancement in controlling tunnel deformations through grouting in mucky soil and soft clay strata, which were previously considered challenging areas for grouting. It has pioneered a paradigm where grouting effects can be anticipated, the grouting process is highly precise and controllable, and grouting disturbances are minimized, making quantifiable grouting possible. The innovative core technology of micro-disturbed grouting (including patented equipment, grouting construction parameters, and grouting processes) can minimize disturbance to the strata during grouting, achieving insignificant disturbance with single hole, single-time grouting operations. With the key technical standard of "evenness, multipoint, small quantity, multiple times", multiple-hole, multiple-time micro-disturbed grouting is carried out, superimposing the effects to reinforce the soft strata under tunnel linings, allowing for a gradual and organized uplift of the settlement zone of the tunnel, adjustment of the tunnel settlement alignment, and effective control of tunnel settlement. Regarding convergence control, the implementation of lateral micro-disturbed grouting on both sides of the tunnel increases lateral resistance and lateral earth pressure. This can rapidly curtail the worsening of tunnel convergence, leading to improvements and stabilization of tunnel convergence.

Compared to traditional grouting methods, the micro-disturbed grouting method focuses on adjusting the tunnel's longitudinal alignment and stabilizing settlement. Its distinct technical features include: The method aims to "adjust and maintain the tunnel's longitudinal alignment"; its main process involves "settlement prediction, zoning and staged grouting" to mitigate or control differential settlement; given the technical requirements, the grouting must be conducted "evenly, in small quantity, multipoint and by multiple times. Each single-time grouting should be carried out in single hole so that multiple grouting operations can achieve the desired effect; the grouting is supported by settlement prediction and information-based construction; in a linear control system, several factors decide the grouting method, including "time, location, sequence, and frequency". Its technical core lies in equipment, processes and construction parameters; and information-based construction is employed for flexible scheduling throughout the whole process. This technology seamlessly integrates deformation prediction and on-site grouting control, offering strong maneuverability during the grouting process, precise estimation of grouting effects, minimal grouting process disturbance, and versatile grouting equipment. It has innovatively achieved remediation of tunnel deformations without disrupting normal subway operations.

Micro-disturbed grouting can address grouting reinforcement challenges during both tunnel construction and operation. Furthermore, it can serve as a medium and long-term technical solution to tunnel deformations. It is widely applied in sections of new and operational lines with frequent issues, such as interconnection passage, pumping stations, portal shafts, crossing sections, areas where incidents occurred during the construction phase, and segments affected by construction activities during operation. Additionally, it is promising for broader application in other domains, including similar underground or ground infrastructures, or building reinforcement, particularly for controlling differential settlement of tunnels in soft strata of coastal cities.

References

Yan J, Wang R (2018) Cause analysis and typical characteristics of settlement and lateral convergence deformation of Shanghai soft soil subway tunnel. J Nat Disasters 27(4):178–187

Zhang D, Zou W, Yan J (2014) Study on lateral grouting control mechanism for large lateral deformation of soft soil shield tunnel. Chinese J Geotech Eng 36(12):2203–2212

Chapter 7
A Lifecycle System of Intelligent Operation and Maintenance Management for Tunnels

With the rapid development of urban rail transit across the country, the maintenance and repair management tasks of urban rail tunnel structures in various regions are becoming increasingly demanding. Therefore, it is urgent to strengthen the informatization management of tunnel structures, establish an intelligent management platform for tunnel structural safety, and address the interconnection and interoperability challenges between tunnel structural basic information, inspections, monitoring, and maintenance information. This will provide strong data support for the daily maintenance management, emergency management, as well as tunnel state prediction, warning, and forecasting of tunnel structures in urban rail transit. As the informatization construction of the Shanghai Metro continues to deepen, there have been many explorations in the field of informatization and intelligent construction for the safety of subway structures. This has resolved information isolation in the information management of tunnels and established a Lifecycle System Platform of Intelligent Operation and Maintenance Management for Tunnels. This platform makes the maintenance of tunnels transparent and allows for real-time monitoring of the operational performance of subway tunnel structures. The platform system mainly includes a Geological Information Management and Analysis System for Tunnels, Monitoring and Measurement Data Management System for Tunnels, Ground Inspection Management System for Tunnels, and Structural Engineering Construction and Maintenance Integration System for Tunnels.

Starting from the demand for intelligent data management, safety assessment, and warning throughout the entire lifecycle of tunnel structures, this chapter focuses on introducing the architecture design, key technologies, and core functions of multiple systems developed and applied in the Shanghai Metro.

R. Wang et al., *Structural Safety Early Warning and Treatment Technology of Metro Tunnel in Soft Soils*, Key Technologies for Tunnel Construction under Complex Geological and Environmental Conditions,
https://doi.org/10.1007/978-981-96-6410-8_7

7.1 Geological Information Management and Analysis System for Tunnels

7.1.1 Overview

Shanghai is a typical soft soil area. The sediments are 200–300 m thick above bedrock, and saturated soft clays dominate within 30 m below the surface. Due to the high compressibility and rheological properties of such soft clays, even after the subway was opened to traffic, the soft strata will continue to experience consolidation and secondary consolidation settlement over a considerable period of time. This characteristic of soft strata poses numerous structural problems for long-term operation of subway tunnels. Therefore, the structural health of subway tunnels is closely related to the surrounding soil and water conditions. In order to effectively manage geological data along the rail transit lines, facilitate efficient retrieval and querying of geological data, a Shanghai operational subway geological database has been developed, which stores a large amount of foundational data and includes practical geotechnical analysis functions.

7.1.2 System Structure

The database system consists of four independent platforms. The overall framework of the system is shown in Fig. 7.1. It includes foundational data such as large-scale vector topographic maps of Shanghai, Shanghai rail transit route maps, and geotechnical engineering data including stratigraphic and engineering geological drilling information, and some subway structural data.

7.1.3 Database Organization and Management

Based on the system's basic functions and data coding, a database management system has been established using SQL Server. The objects of data model in the database involve information tables of tunnels geological engineering, drilling, stratigraphic section, strata, drilling strata, and subway structure. These efforts are made to create a geospatial property database and associate graphic property with index fields. The graphic property values are used to locate database index fields, which are then used to retrieve other related field contents.

For various types of spatial entity objects, CAD original graphic data is first standardized into the Shanghai urban coordinate system. Then, based on the coordinates of each entity object, precise positioning is carried out to determine the

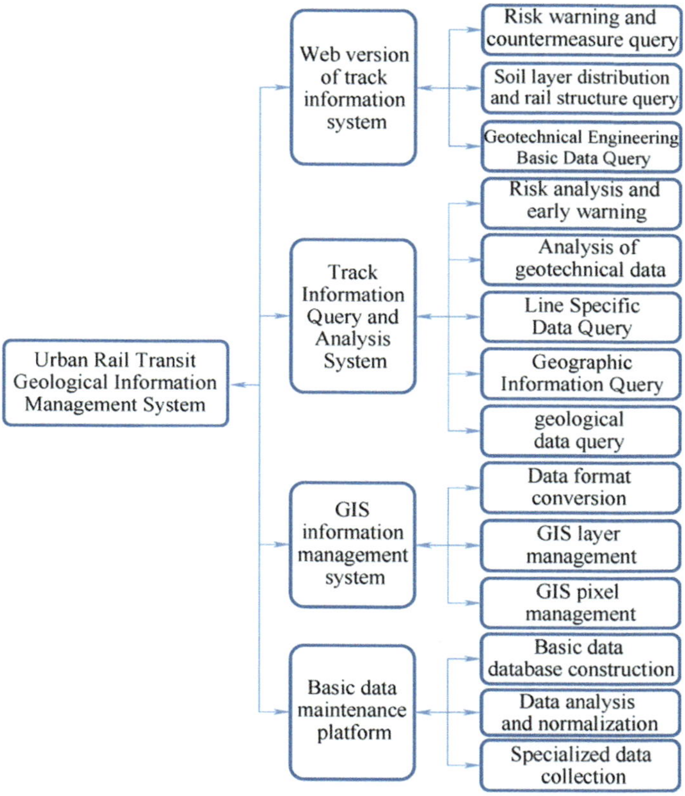

Fig. 7.1 Overall framework of the system

location relationship of topography, tunnels route directions, along with geological engineering, and drilling distributions. This is achieved through layer-by-layer management, enabling the display of various foundational data for tunnels.

During the system implementation process, a systematic process was established from platform construction, data standardization, data collection, data input, data query, to statistical functions development. Finally, with the help of an expert team, specific geotechnical analysis was conducted on geological data.

7.1.4 System Functions and Technical Highlights

A GIS-based tunnels geological database platform has been constructed, enabling spatial data storage and display. It has established an efficient indexing mechanism which provides practical real-time geological data query and statistical functions.

The platform fully realizes the digitization and visualization of geological data and has developed geotechnical analysis functionality based on geological data.

7.2 Monitoring and Measurement Data Management System for Tunnels

7.2.1 System Overview

Soft soil subway tunnels are susceptible to various factors such as construction quality, surrounding construction projects, and complex geological conditions, which lead to a relatively high incidence of structural diseases. To address these, a large number of monitoring and testing measures are required, and the resulting data needs to be managed informatively to enhance control abilities and efficiency. In response to the challenges of extracting, storing, cleaning, and utilizing massive and heterogeneous data from structural monitoring and measurements in soft soil tunnel structures, based on the national "Tianditu" platform, as well as Internet of Things and cloud technologies, the Monitoring and Measurement Data Management System for Tunnels was created, which is unprecedented in China.

The management system supports the integration and application of stereo perception data, including satellite images, low-altitude remote sensing images, mobile panoramic photography, real-time monitoring with automated sensors, and precise engineering measurements. It achieves the integration and fusion of "air + ground + underground" remote sensing images, laser scanners, sensors, and precise engineering measurements, as shown in Fig. 7.2. Through data analysis and mining, it conducts structural safety assessments and risk management. It also provides convenient and reliable data management and engineering information services to the tunnel operation and maintenance industry by enhancing data acquisition capabilities, innovating service modes, improving operational efficiency, and supporting risk control, therefore comprehensively and efficiently controlling the impact of surrounding construction activities on subway structural safety.

7.2.2 Perception Methods and Data Integration

Stereo perception aims to integrate information sources from air (satellite, UAVs), ground (panoramic images, manual inspections), and underground (precise engineering measurements, automated sensors, mobile laser scanners, etc.). The integration achieves standardization and high efficiency in data interface, visual interface, and information integration, processing and analysis, expanding the understanding of professionals about structural safety monitoring data, as shown in Fig. 7.3.

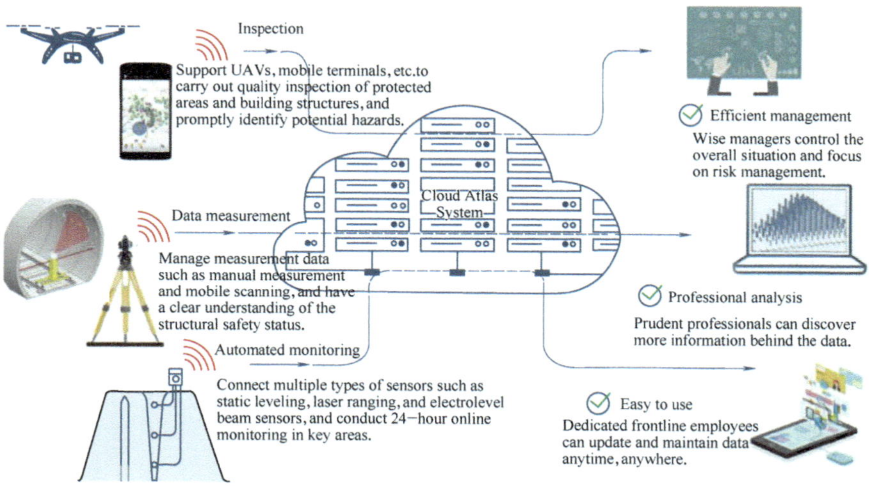

Fig. 7.2 Tunnel supervisory measurement data management system

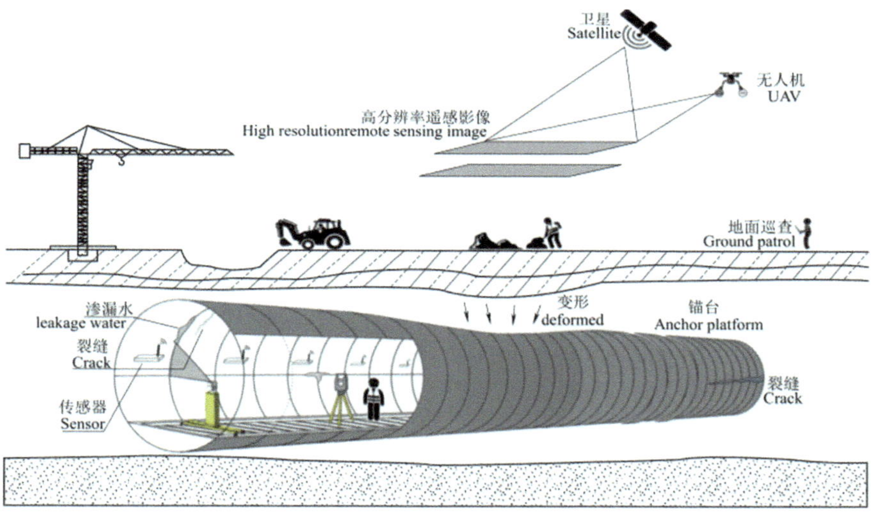

Fig. 7.3 Conceptual diagram of stereo perception of tunnel safety

7.2.2.1 Perception Technology Means

Based on the location and purpose of data collection platforms, data for tunnel protection zones and structural safety can be collected through air, ground, and underground channels. These methods are primarily used for detecting unauthorized activities

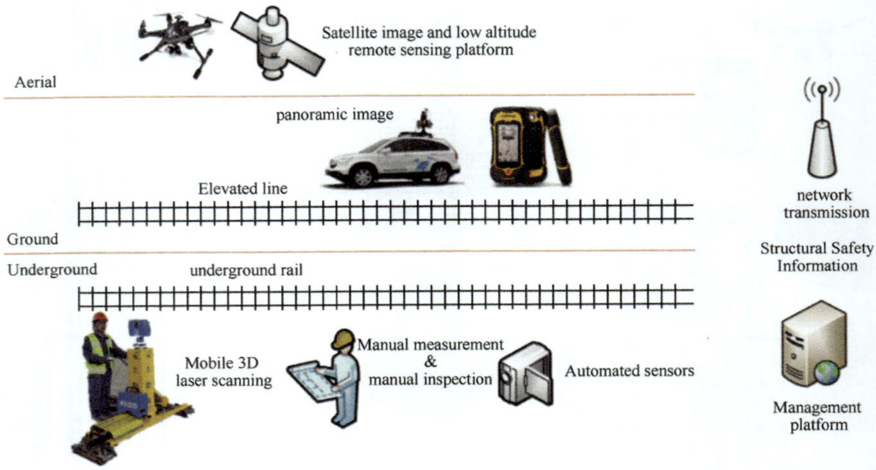

Fig. 7.4 Data acquisition methods ("air ground underground")

within the protection zone, monitoring long-term deformation patterns, and supervising engineering impacts. The perception methods are illustrated in Fig. 7.4 and listed in Table 7.1.

7.2.2.2 Data Integration

The standardized "air + ground + underground" data management aims to reduce errors in data entry and information gaps. It utilizes data maintenance tools to achieve standardized retrieval after initial entry, bringing in optimal organization and management efficiency. The requirements for data sources and data formats can be found in Table 7.2, and the data application requirements are listed in Table 7.3.

To ensure effective information transmission, the basic requirements for data transmission are as follows:

(1) Procedural collaboration: After the raw data is collected and processed, they undergo standardized review and risk analysis;
(2) Whole process recording: Entire process recording is beneficial for daily management, subsequent summaries, improvement, and tracking of key risks;
(3) No transmission failure: Avoid changes or losses of data and documents during transmission;
(4) Access anytime, anywhere: Implement information retrieval, updates, and maintenance services in different locations, across various media, and with controlled permissions.

Table 7.1 Perception objects, means, application requirements, and objectives

Perspective	Objects	Means	Application requirements	Objectives
Aerial	Security protection zone	Satellites and low-altitude remote sensing	Ground inspection	Detect ground activities and conditions that may impact structural safety
	Elevated structures	Low-altitude remote sensing	Structural deformation monitoring	Monitor deformation of elevated structures
Ground	Security protection zone	Panoramic images, manual inspection	Ground inspection	Detect ground activities and conditions that may impact structural safety
	Elevated structures	Precise engineering measurement, automated monitoring	Structural deformation monitoring	Monitor deformation of elevated structures
Underground	Tunnel and underground station structures	Laser scanner, precise engineering measurement, automated sensors	Structural deformation monitoring	Monitor deformation of tunnel structures
	Tunnel and underground station structures	Laser scanner, infrared imaging	Structural disease assessment	Survey tunnel structural disease
	Tunnel and underground station structures	Manual inspection	Structural disease assessment	Conduct detailed inspection of tunnel structural disease

7.2.3 Architecture Design

7.2.3.1 Overall Design Principles

The overall design principles of the database system are mainly:

(1) Practicality: Demand-oriented, able to balance the comprehensive functions and performance of the system.
(2) Security: Strictly adhere to national and corporate data confidentiality standards, prevent illegal intrusion, and encrypt network transmission to avoid information loss or tampering.
(3) Reliability: Implement debugging and log tracking functions, automatically and timely recover in case of breakdown, and quickly resume operation. The system should also have data-level disaster recovery function and achieve real-time backup in different locations.

Table 7.2 Data sources and data formats

Item	Data source	Data format	Order
Inspection	Center line, edge line	3D coordinates (points, lines, polygons)	< 1 Gb
	Mileage (ring number)	Mileage stake and ring number	< 1 Gb
	Topographic map	Topographic map along the route	< 100 Gb
	Image data	Ground images, satellite imagery, UAV aerial photography, etc.	> 100 Gb
Deformation	Settlement	Height	< 10 Gb
	Converge	Measured value, design value	< 10 Gb
	Lateral displacement	Displacement value	< 10 Gb
	Titling	Titling amount and slope	< 10 Gb
	Crack	Opening displacement	< 10 Gb
	Joint dislocation	Joint dislocation value	< 10 Gb
	Layered settlement of soil	The amount of layered settlement	< 1 Gb
Disease	Image	Ordinary camera, laser scanner, linear array photography	> 10 Tb
	Recognition result	Disease type, location, area, etc.	< 1 Gb
	Statistical data	Classification and time statistics results	< 1 Gb
Archives	Business data	Review process, official correspondence, information notifications, etc.	< 5 Tb
	Technical files	Drawing documents, technical reports, etc.	< 5 Tb
	Process documents	Meeting summary, report materials, stage summaries, etc.	< 5 Tb
	Others	Archive, backup, and other materials	< 10 Tb

(4) Scalability: Develop based on domain-oriented interfaces, achieve low coupling of modules, reserve sufficient upgrade interfaces and redundancy space, reserving better openness and maintainability.
(5) Advancement: Use internationally recognized advanced technologies that align with the system's characteristics to improve development and operational efficiency.

Table 7.3 Content and application requirements for data sharing and information transmission

Types of application	Content	Requirements of application
Data sharing	Line section, structural form, station form, etc.	Centralized management
	Information, such as structural location and the scope of the protected area (partially classified)	Basic surveying and mapping data
	Office plugins, online input tools	Data maintenance tools
	Standard properties or data for easy database management	Structural data
	Drawings, images, etc. saved in file format	Files data
	Office online, Autodesk online viewer, etc.	Extended services
Message passing	Drawings, reports, etc. formed during the engineering project process	Current situation map data
	Real-time acquisition and visualization	Real-time monitoring data
	Query, statistics, and visualization of structured data, archiving of paper (electronic) materials, etc.	Historical data
	E-mail, QQ, WeChat, SMS, etc.	Instant messaging
	Business flow or information transmission, exchange, etc.	Process transmission
	Retention of records and materials formed during meetings and other processes	Others

7.2.3.2 Overall Architecture Design

The system is divided from top to bottom into presentation layer, application service layer, domain layer, and infrastructure layer, as shown in Fig. 7.5.

In the field of tunnel engineering, the presentation layer is implemented in a front-end Web interface. It is responsible for managing user-system logical interactions and enabling the publishment, display and management of information resources.

To achieve better decoupling and modularization, the system further divides the business logic layer into application service layer and domain layer. The former focuses on handling users' business operations, isolating the presentation layer and domain layer, enabling effective code reuse and logic adjustments, and providing greater flexibility.

The domain layer is the core structure of the system. It abstracts and models the business, forming an object model and operational logic that reflects the domain.

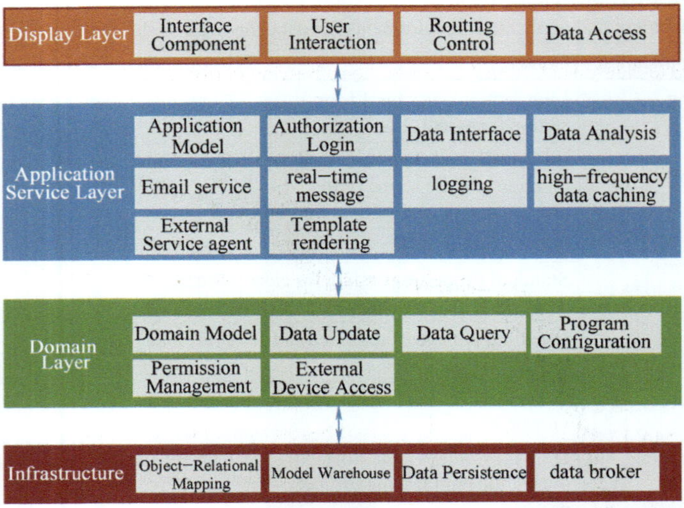

Fig. 7.5 Overall system architecture design diagram

The infrastructure layer is the lowest level of the system. It provides practical support for the domain layer and application layer business services through various components, including data persistence and data backup.

7.2.3.3 Server Architecture Design

As a multi-source data integration platform, the system includes a public cloud network and a real-time transmission network based on the Web server, as shown in Fig. 7.6.

Figure 7.6 describes the composition of the network architecture of this system. The application servers are the core processing units of the business, consisting of one to multiple server clusters surrounded by supporting servers that handle various data and business operations. Supporting servers are an abstract concept which can be deployed either in a distributed manner or on the same physical machine. And the servers are logically isolated from each other and only interact with application servers through interfaces or APIs. Therefore, such servers are mutually independent, making it convenient to horizontally scale and allocate loads as the system's application scope continues to extend.

7.2.3.4 Front-End and Back-End Design

In the front-end, the system utilizes the MVVM framework to build single-page, highly interactive Web applications. Leveraging the advantages of two-way data

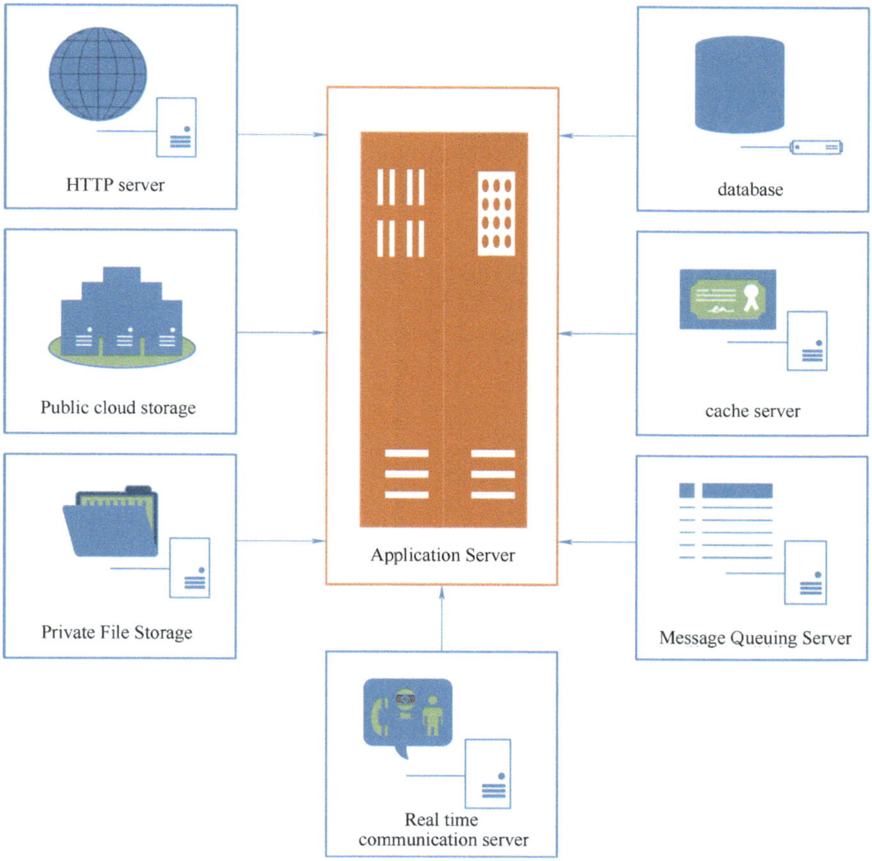

Fig. 7.6 System network framework

binding brought by MVVM, the program can separate the view layer, enabling the development of a single set of models (M) and view models (VM) with multiple views. This approach supports devices in sizes, enhances collaborative efficiency, and contributes to the delivery of more reliable code.

In the back-end, the system employs Node.js and ASP.NET for development. Node.js handles the logic services at the application layer and view rendering, while ASP.NET focuses on domain modeling, providing data support at the business layer. By using Node.js and ASP.NET for back-end development, a natural separation and distinction between the application and domain layers are achieved, which is advantageous for future system expansion and optimization.

For mobile devices, as both Node.js and the front-end are written in JavaScript, the system adopts a Hybrid HTML development approach, reusing a significant amount of code. Additionally, the REST API on the Node.js side serves data interactions within the hybrid app, eliminating the need for additional back-end code. This hybrid development approach greatly shortens the development time for mobile apps. The final app supports both Android and IOS platforms, reaping significant Return on Investment, ROI.

7.2.3.5 Information Security

Information security must meet the first-level requirements. The main components include: (1) Identity authentication. (2) Access management. (3) Security audit. (4) Intrusion prevention. (5) Malicious code prevention. (6) Resource control. (7) Data confidentiality and recovery design.

7.2.4 *Function Implementation*

The overall system functions are illustrated in Fig. 7.7. Through tunnel monitoring methods and the integration of a cloud-based monitoring information system, the system ultimately achieves risk management for tunnels.

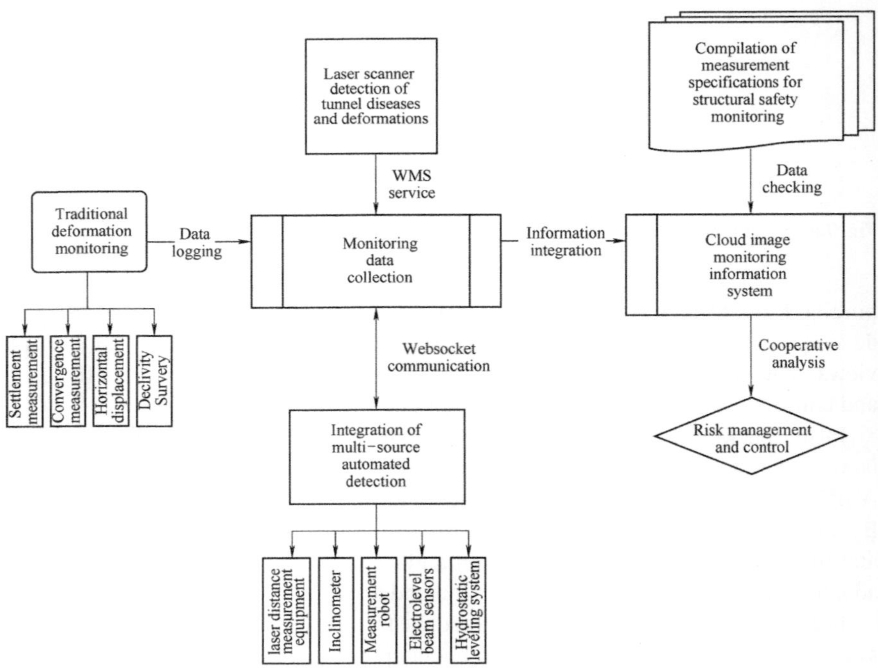

Fig. 7.7 Overall system functions

7.2.4.1 Sensor Management Module

(1) Module architecture

The sensor management module within the platform is responsible for controlling all transactions related to automated sensors within the tunnel. They include:

(1) Periodically collecting data according to predefined schedules.
(2) Receiving measurement commands from users and collecting real-time data.
(3) Offering version updates, configuration synchronization, data uploads, and data synchronization.

The physical architecture of the sensor management module consists of sensors, IPCs, mobile network routers, servers, and a website as illustrated in Fig. 7.8.

(2) Sensors

Sensors are devices designed to directly capture physical measurements. They are typically installed within the tunnel and connected to the IPCs via RS-232 or RS-485, as depicted in Fig. 7.9.

Sensors vary slightly in connection ways. Hydrostatic leveling systems, electrolevel beam sensors, and laser range finders are parallelly connected, while for Leica total stations, one USB data cable is connected to one sensor.

(3) Industrial personal computer (IPC)

As shown in Fig. 7.10, the IPC will automatically start upon booting up. It reads the data from pre-configured automatic observation systems, synchronizes the latest measurement configurations with the server, and starts three threads to perform these three tasks:

① Set up persistent connections with the server and establish a real-time communication channel to quickly respond to timely measurement commands from users.

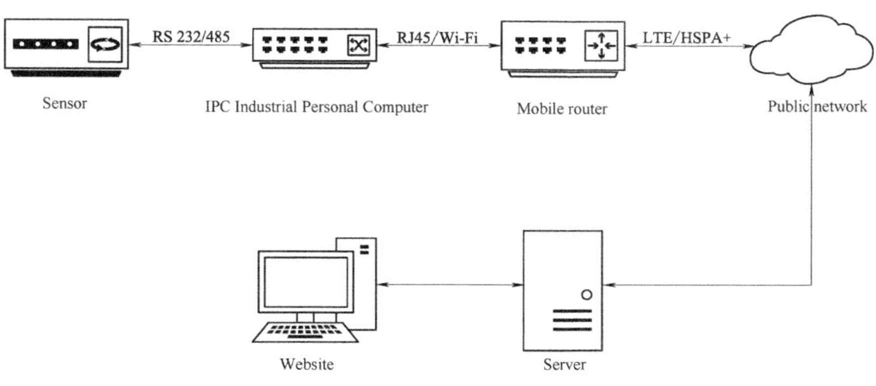

Fig. 7.8 Physical architecture of the sensor management module

(a) Total Station (b) Hydrostatic Leveling System (Hls) (c) Inclinometer

(d) Electrolevel Beam Sensors (e) Laser Distance Measurement Equipment

Fig. 7.9 Sensors

Fig. 7.10 IPC sample

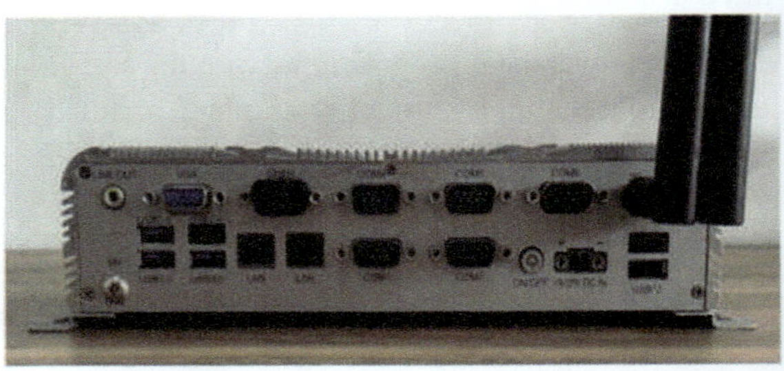

② Implement a timely measurement program to automatically measure data at the specified time.
③ Check for program updates. The IPC will automatically download the latest version and update the program.
④ Persistent connections and a real-time measurement protocol.

The real-time measurement of the sensor management module relies on persistent connections and a self-developed real-time measurement protocol.

Persistent connections are regulated by the transmission control protocol (TCP), an industry-standard protocol suit. When incorporated in the module, the connection technology is slightly different from the common practice. Each automatic observation system of IPC, the client, will first establish TCP connections with servers; then it sends unique identifiers (UID) to servers, finishing a sign-up process; and it will timely send heartbeat packets to check the connection status. The establishment of persistent connections helps create a two-way channel between the IPC and the server. Therefore, the channel achieves data transmission between them at all times, laying the foundation for real-time measurement.

Based on persistent connections, a set of real-time measurement protocols are required to specify the details of real-time measurement, including instructions for starting, stopping, refreshing, returning, etc. And the timely measurement protocol also involves several instructions, as listed in Table 7.4.

(5) Data synchronization technology

Due to weak stability and transmission speed of mobile networks in long tunnels, persistent connections often suffer from disruptions, leading to data loss. Therefore, the sensor management module employs data synchronization technologies by building a local database on the IPC to temporarily store the configurations and measurement results of automatic observation systems. Supported by stable wireless networks, measurement results are uploaded to the server, and the latest configurations of automatic observation systems are downloaded from the server. Only when data synchronization is successful, can the copies in local database be deleted.

(6) Server

To establish communication channels between the website and the server, as well as between the IPC and the server, the sensor management module deploys Redis-based message publishing and subscribe service functions on the server to exchange information between both parties. As the network hub between users and IPC, the server also acts as the data hub for all automation projects to store related configurations and measurement results.

(7) Data processing

A data processing sub-module is deployed on the server to process raw measurement data, as outlined in Table 7.5.

(8) Website

According to the logical structure of the sensor module, the website is regulated at different levels, creating a general interface and individual ones. Each individual interface is subdivided into sectors, involving measurements, layout maps, data queries, configurations, diagrams and log queries. The Angular front-end framework can efficiently integrate the content, style, and logic involved in pages into independent modules.

Table 7.4 Periodic survey protocol

Type of command	Name of command	Code of command	Meaning of command
Server commands	Start survey	StartSurvey	The server sends a StartSurvey command to the IPC
	Stop survey	StopSurvey	The server sends a StopSurvey command to the IPC
	Reload config	ReloadConfig	The server sends a ReloadConfig command to the IPC
	Report log	ReportLog	The server requests a report log from the industrial computer
IPC commands	Register automated stations	Register	IPC registers automated measuring stations with servers
	Point survey result	PointResult	IPC reports the measurement results of a single measurement point to the server
	Point survey failure	PointFailure	IPC reports the failure of a single measurement point to the server
	Start survey	StartSurvey	IPC reports to the server that this measurement has started
	Survey complete	SurveyComplete	IPC reports to the server that this measurement has finished
	Reload config	ReloadConfig	IPC reports to the server that the configuration has been updated
	Report log	ReportLog	IPC reports logs to the server
	Plain text message	Message	The IPC sends messages to the server

(9) WebSocket Technology

In addition to traditional HTTP, the communication mechanism between the website and the server also utilizes WebSocket, an emerging technology. The technology offers two-way real-time communication capabilities, reducing the workload for developers and efficiently achieving real-time communication. The sensor management website employs WebSocket technology, using a self-developed real-time measurement protocol to enable real-time measurement.

Table 7.5 Server data processing module table

Types of sensor	Raw data	Data processing method	Final outcome
Distance measurement equipment	Horizontal diameter of tunnel	Gross error elimination	Horizontal diameter convergence of tunnel
Inclinometer	Tunnel segment inclination angle	Gross error elimination	Tunnel segment inclination angle
Hydrostatic leveling system	Level height	Gross error elimination Height transmission	Vertical displacement of tunnel
Electrolevel beam sensors	Voltage	Gross error elimination	Vertical displacement of tunnel
Total station	Horizontal angle, zenith distance, slant range	Gross error elimination Combined adjustment of multiple stations	3D coordinates of each measuring point

7.2.4.2 Data Maintenance Plugin

Considering that operators frequently use Excel to record manual measurement data and compile monitoring reports, an Excel-based data maintenance plugin has been developed to facilitate the upload of manual measurement data.

The Excel plugin supports the .NET Framework and can use its class library featuring rich resources as well as third-party libraries to design a wide range of functions. The plugin fully integrates with Excel, helping operators easily get started. It also supports automatic updates and has a compact size of around 1 MB, ensuring a good user experience and quick updates.

7.2.4.3 Statistical Analysis Interface

The platform provides the function of statistical analysis for data testing. Due to the large data sets involved in the entire line and long computation time, the platform employs background computing techniques to periodically crunch results and cache them in the database. When users access the statistical analysis interface, the platform directly retrieves results from the cache memory, minimizing waiting times.

7.2.4.4 Visual Interface

(1) Interaction logic of the interface

Given the diverse data structures and complex business logic of the platform, Angular and JavaScript are used to implement the interaction logic of the visual interface. The website is divided into several states based on business logic, with each state corresponding to one controller. Each controller will write code related to interface interaction, and invoke data structures for data modification and calls to third-party libraries, to perform functions, like diagram creation and image rendering.

(2) Data visualization

The platform's data can be categorized into two types: spatial information and projects. For both types of data, the platform employs techniques such as distribution maps and line charts to achieve visualization.

In the maps, different colors and individual characters are used to differentiate project levels. Users can easily discern project levels and their relationships with railway lines from the project map, as shown in Fig. 7.11.

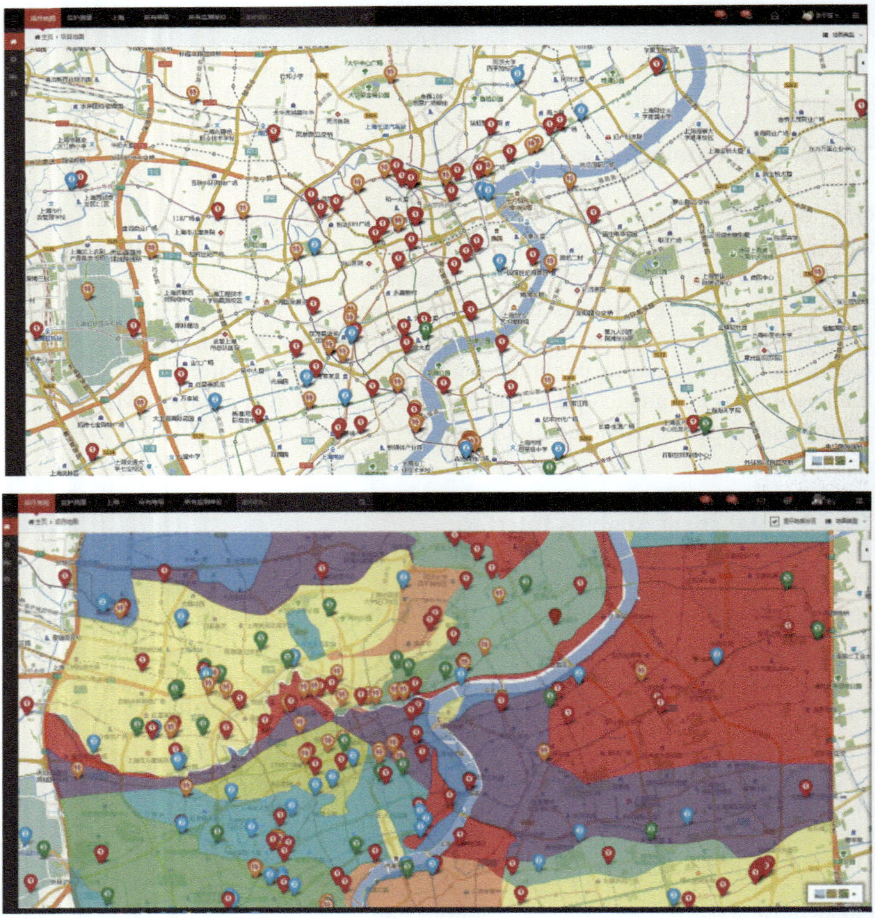

Fig. 7.11 Project map

The monitoring point distribution map displays information about the locations of monitoring points on the track, as well as their distances from the construction area, as shown in Fig. 7.12.

The tunnel monitoring results are represented by line charts, as shown in Fig. 7.13. These charts are used to display the cumulative changes between the current and the previous measurement. The distance between the two lines on the chart represents the changes between the current and previous measurements. The line charts can also be customized to simultaneously represent multiple lines for nearby projects, making it convenient for users to compare and contrast from different aspects and conduct an integrated analysis.

The platform is provided with the interactive operation function to associate the point distribution map and the line chart. When a user hovers the mouse over a specific point on the line chart, the distribution map zooms in on the same point with highlights, as shown in Fig. 7.14.

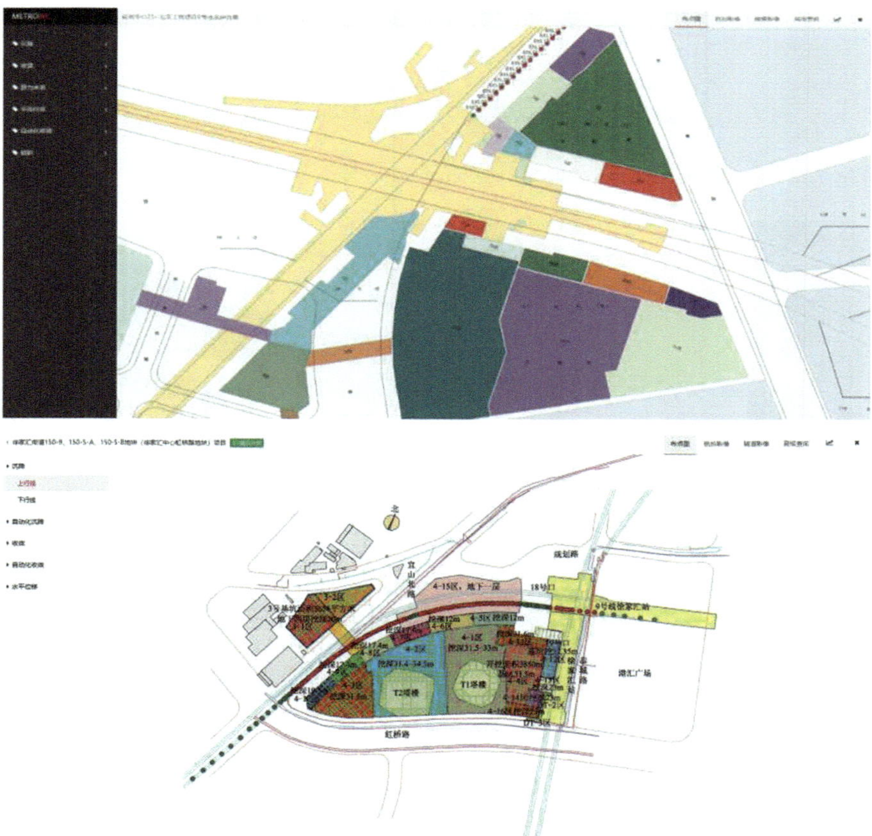

Fig. 7.12 Project monitoring point distribution map

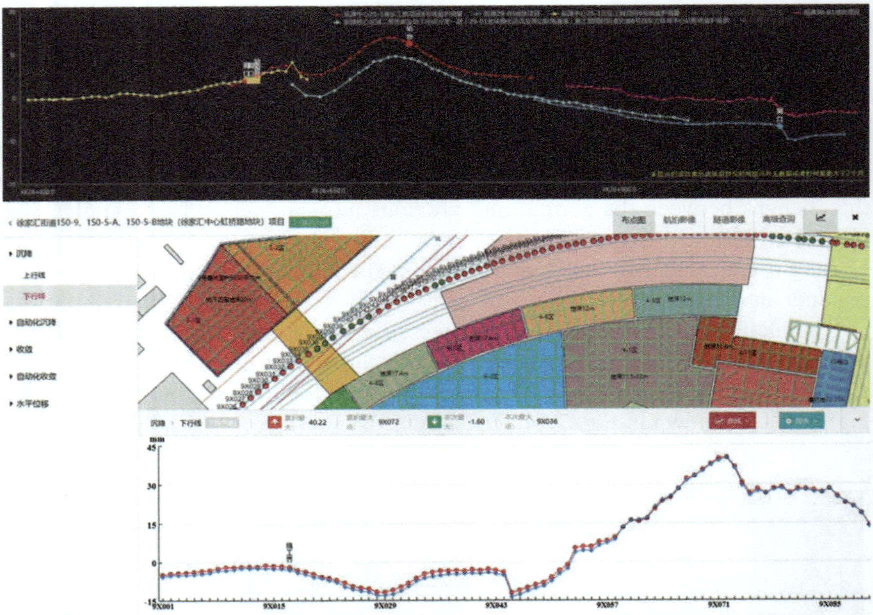

Fig. 7.13 Tunnel monitoring results curve comparison chart

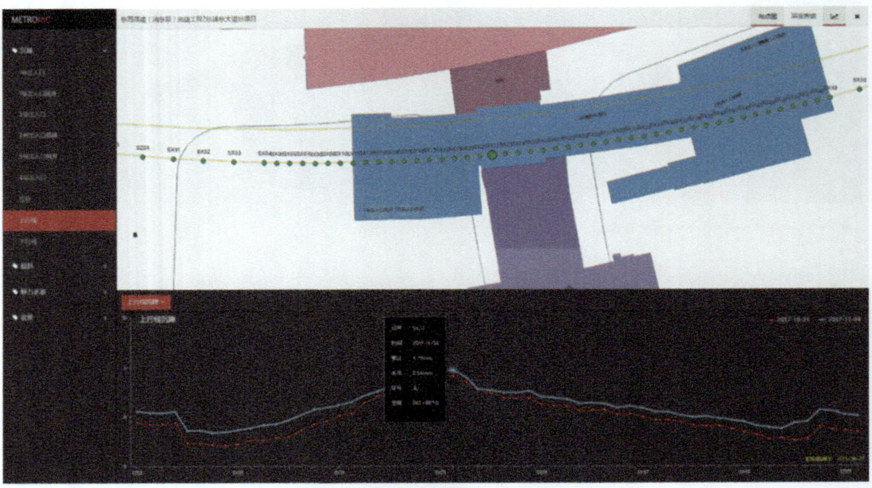

Fig. 7.14 Scatter plot and line chart interactive interface

7.2.5 Application

The platform has been utilized in structural deformation monitoring for the Shanghai Metro. The management data includes measurement, image, and other properties (construction information, photos, and point layout diagrams). The tunnel monitoring projects are categorized into supervisory measurement, long-term settlement measurement, long-term convergence measurement, and key section measurement.

The platform manages an annual increase of 300 projects for supervisory measurement, 181 long-term settlement data intervals, 282 long-term convergence intervals, and 111 key sections. It also publishes and stores three-dimensional laser scanner images and UAV aerial images in different time periods.

7.3 Ground Inspection Management System for Tunnels

In early times, operators conducted inspections for subway protection zones by distributing paper task sheets and by using external positioning devices and digital cameras. However, problems arose, including bulky equipment, poor usability, ineffective information integration, uncontrollable inspection quality, such as missed or wrong inspections. To address these issues, a new inspection management model, "Intelligent Ground Inspection System for Tunnel Protection Areas", was developed for the first time, which incorporated cloud computing, mobile Internet, and GIS. The model combines three factors in one whole, including industrial handheld terminal devices, a mobile operating APP, and a WEB-based project control system.

7.3.1 Hardware Equipment

The system uses GNSS A5-upgrade hardware, a leading technology in China. The A5 upgraded version of the Beidou intelligent terminal is equipped with a 64-bit quad-core 1.2 GHz Snapdragon processor, offering higher working efficiency and ultra-low power consumption based on the Cortex-A53 architecture. The device runs on Android 5.1.1, supporting Beidou, GPS, and GLONASS GNSS satellite reception, as well as TD-LTE/FDD-LTE dual 4G network communication, meeting the needs of big data transmission in an era of mobile Internet. It has a 13-million-pixel camera and supports professional and outstanding photography, suitable for capturing image and video property information, ideal for a solid intelligent terminal in Internet era. The device can access differential signals through Ntrip, enabling differential operations and achieving sub-meter positioning accuracy.

7.3.2 *Software System*

7.3.2.1 Webpage End

The webpage end has four functions: task management, task evaluation and supervision, task results review, and real-time query.

1. Task management

This function involves task planning, distribution, specifications, and approval, forming a comprehensive task management system.

2. Task completion evaluation and supervision

Task completion status is visually displayed in graphical form, represented in terms of the day, cycle time or history for both single lines and the entire network, along with corresponding inspection records.

3. Task results review

Project conditions reported by inspectors require verification and approval by administrators. Click the "View" button to see detailed project information. Approved projects will be included in the inspector's workload.

4. Real-time query

The system shows task query, inspection records, project distributions, and other information on the map. Results can be filtered by inspector, route, date, and other criteria.

7.3.2.2 Mobile End

1. Terminal confidentiality

The terminal employs account-password and hardware serial number verification. In the event of device loss, users can dispose in time and delete the serial number. The system offers two login modes: personal and group. Personal login is used when only one person is required to perform and follow projects or is temporarily assigned to the projects. Group login is used when multiple individuals simultaneously conduct routine inspection, requiring simultaneous verification of respective login usernames and passwords.

2. Task reception

After logging into the system, all tasks for the current inspection cycle are displayed. Clicking on a specific date button reveals detailed task categories, including routine, tracking, and self-created tasks.

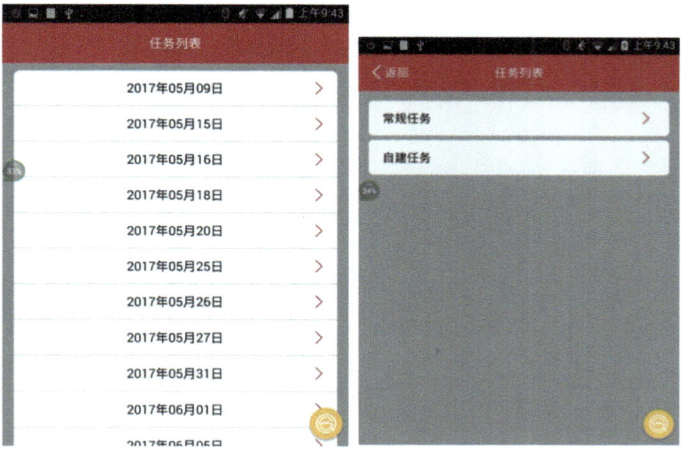

Fig. 7.15 Mobile terminal diagram

3. Inspection and violation reporting

Inspectors can visually view the starting point, current location, endpoint, station names, and subway protection area ranges for the current inspection task using the handheld terminal.

The system provides a project list function for users to directly query all types of project information recorded within their daily task areas. Projects can be directly centered on the map for violation reporting (Fig. 7.15).

When a violation is found, users can input the coordinates and access the project details interface. They can follow the prompts to fill in the basic information of the project and report violation information by taking pictures or making tape recordings.

4. Track uploading

As inspection begins, the movement trajectory is uploaded to the server at a rate of one point every 20 s.

7.3.3 System Innovations and Application Effects

7.3.3.1 Establishment of a New Mobile Inspection Mode with Satellite Precise Positioning

The new mode is to help inspectors access specific locations of subway underground facilities and the spatial relationships between construction projects and facilities. It adopts a mobile terminal equipped with Beidou, GPS, and GloNass GNSS reception.

Fig. 7.16 Real-time display of location information on handheld terminal inspection

Supported by GNSS positioning, based on urban topographic and tunnel route maps, the mode displays real-time information on the location of tunnels and the protection areas in the inspection terminal APP. The positioning accuracy reaches sub-meter levels, as shown in Fig. 7.16. Real-time feedback on the distance between personnel's location and the structural boundary enables tracking of personnel trajectories, guiding inspectors in conducting detailed inspections.

7.3.3.2 Efficient and Standardized Collection of Violation Information Along the Line

A handheld terminal APP was developed to assist inspectors in reporting information, standardizing the content of violation reports, and decreasing the operating difficulty. It not only records all inspection process details but also provides user-friendly functions such as photo capture, voice recording, dynamic detection of violation ranges, and construction signs recognition. This establishes the foundation for paperless and process-oriented management. The main functions of the mobile APP are shown in Fig. 7.17.

7.3.3.3 Development of a GIS-Based Inspection and Control Platform

Based on WEB GIS, an inspection and control platform was built to comprehensively integrate information on safety inspections and monitoring of subway lines. This platform fully digitizes and eliminates paper from the entire task management process, including task formulation, allocation, specifications, review, and approval. It also realizes data visualization by which users can access the real-time locations of inspectors and progress of the current inspections, ensuring dynamic and precise personnel management, as shown in Fig. 7.18.

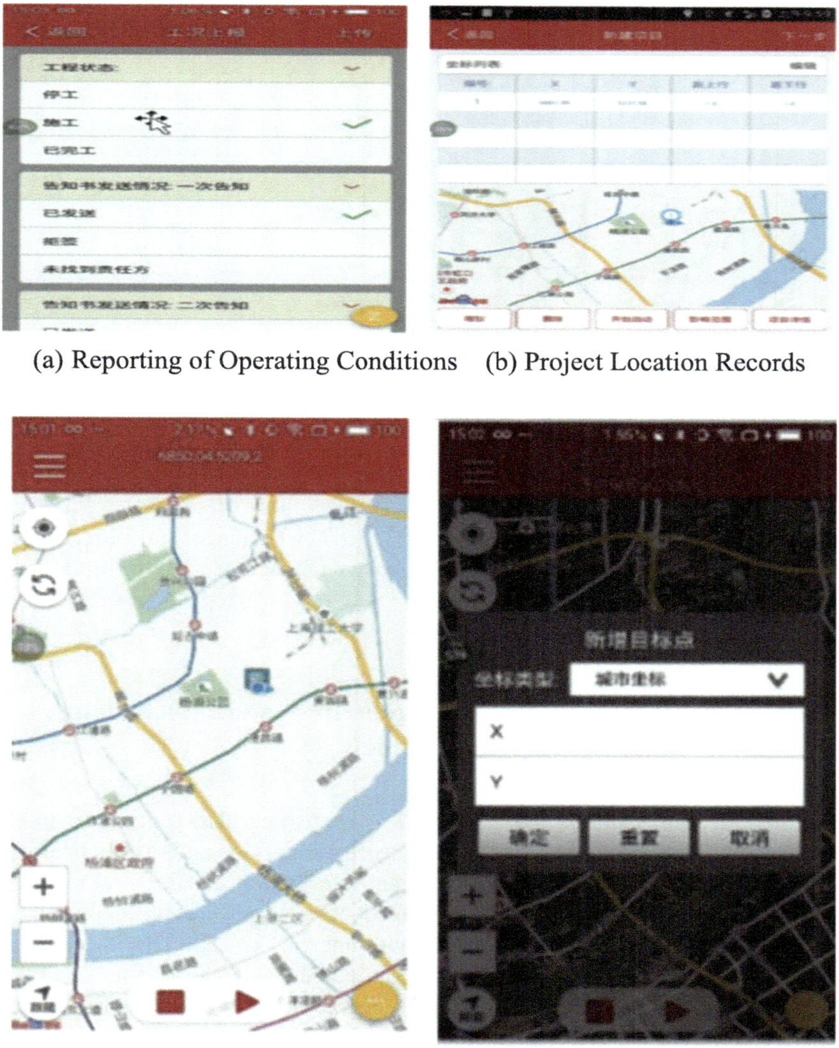

(a) Reporting of Operating Conditions (b) Project Location Records

(c) Personnel Tracking Records (d)Coordinate Marking of Impact Area

Fig. 7.17 Main functions of mobile terminal

7.3.3.4 Establishment of a Collaboration Model of Problem Solving

Leveraging Internet thinking and continuously improving the management model through ongoing research and development, more than 40 versions have been iterated to create a relatively mature and stable system product. Multiple training sessions and technical competitions have enhanced the skill proficiency of operators. Currently,

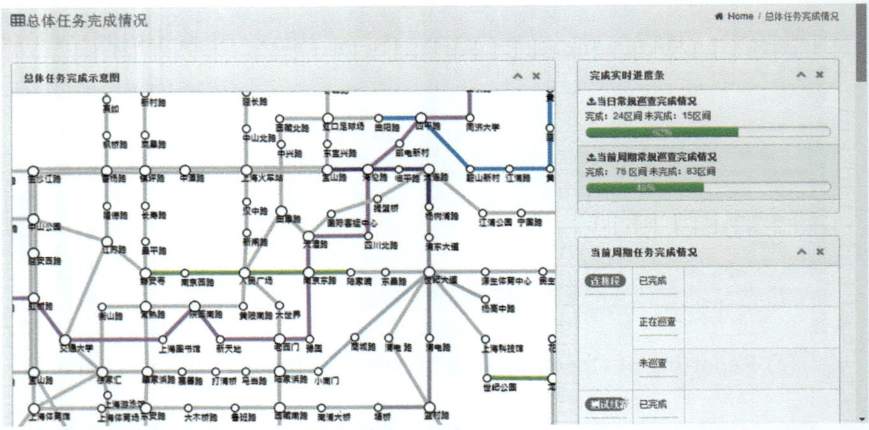

Fig. 7.18 Platform control interface

the system is continuously upgraded to meet new management requirements, in line with the construction of the comprehensive management platform for Shanghai Metro operations.

7.4 Digital Information Management System for Urban Rail Transit Operating Tunnels

As the structural maintenance of tunnels is ever important, it is urgent to strictly control construction quality during the construction period and to effectively collect data during this phase. Currently, important information on structural maintenance during the construction phase is not effectively centralized and analyzed, posing great challenges to safety control and environmental protection during construction and causing significant difficulties in operation and maintenance. The "Digital Information Management System for Tunnel Operating Tunnels" is built centering on integrated construction and maintenance.

The system first systematically categorizes and organizes various types of information during the construction phase, reaching efficient management using database technology. It utilizes network and visualization technologies for simple interactive queries, followed by timely data analysis and efficient utilization. When tunnel construction is completed, it offers structural maintenance and safety management through a 3D digital model and electronic data integrated system. Additionally, the system will integrate a state-based dynamic maintenance planning system and provide data interfaces for automatic monitoring during the operational phase, facilitating seamless integration into the system and demonstrating the full lifecycle digital model.

7.4.1 Functions of the Digital Information Management System

The functions of the digital information management system for urban operating tunnels are illustrated in Fig. 7.19. Basic functions encompass:

(1) Standardization technology: The standardization of critical information and data throughout the entire lifecycle of subway engineering, including investigation, design, construction, monitoring, surrounding environments, and unforeseen events, as well as operational maintenance and monitoring data.
(2) Data compilation and classification: Organizing and categorizing various data types related to investigation, design, construction, monitoring, and surrounding environments during the construction phase. This involves standardizing data based on intrinsic patterns while aligning with national standards, industry norms, and engineering practices. Furthermore, it includes the logical and physical design of a comprehensive database covering the entire lifecycle of tunnel engineering. This encompasses the establishment of naming conventions and metadata for database tables, the research and development of logical access models and unified interfaces, and the formulation of database security policies and sharing mechanisms.
(3) Integration of spatial data: Utilizing spatial database techniques, based on target objects such as tunnels and geological layers within the digital model. This involves associating spatial location and property data to integrate various property data related to design, construction, monitoring, and operational maintenance. The target is to create an efficient lifecycle digital information sharing platform of tunnel engineering.
(4) Evaluation and archiving: Utilizing existing monitoring and testing technologies, alongside various types of information gathered during the processes, to

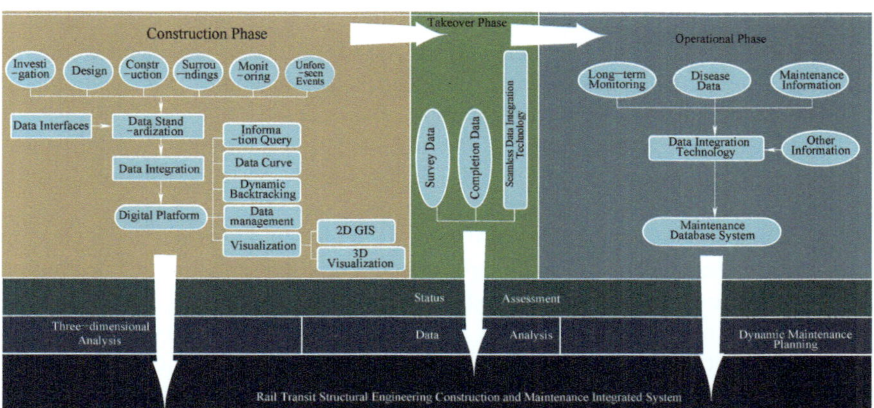

Fig. 7.19 Functionality of digital information management system for urban operation tunnels

evaluate different phases in the lifecycle of tunnel engineering. The results of these evaluations are quantified and archived to serve management decisions and operational maintenance tasks.

(5) Long-term monitoring and maintenance: Creating plans for long-term monitoring in consideration of data continuity and traceability. This process employs data interfaces and data fusion techniques to seamlessly transition various types of information from the construction phase of structural engineering to the operational phase.

(6) Digital management and visualization: Employ computer, database, and digitalization technologies to manage data throughout the entire lifecycle of tunnel engineering, displaying in the 3D digital model.

7.4.2 System Functional Framework

Based on the system's purposes, along with research findings on potential user information needs and work distribution, the system can be divided into several functional modules, including comprehensive information, monitoring and supervision, disease management, maintenance management, status assessment, virtual simulation, and access management.

The comprehensive information module primarily focuses on the management and presentation of investigation, surrounding environment, design, and construction information.

The monitoring and supervision module emphasizes the management, querying, and representation of long-term settlement and section convergence data for post-operation structural monitoring.

The disease management module deals with the uploading, management, querying, statistics, and display of tunnel disease information during the operational phase.

The maintenance management module focuses on the management, querying, and display of historical maintenance information for tunnels. Three factors, monitoring and supervision data, disease and historical maintenance information, are fundamental parameters for tunnel condition assessment.

The status assessment module has an embedded algorithm in the system, which allows for manual intervention through the interface to conduct individual status assessments and display the assessment results.

The virtual simulation module constructs 3D models and provides a VR viewing experience. It enables the management, querying, and presentation of information across the entire lifecycle of tunnel engineering.

The access management module, through user-specific permissions settings, allows for functional personalization, ensuring information security management and system usability.

The system's functional framework is illustrated in Fig. 7.20.

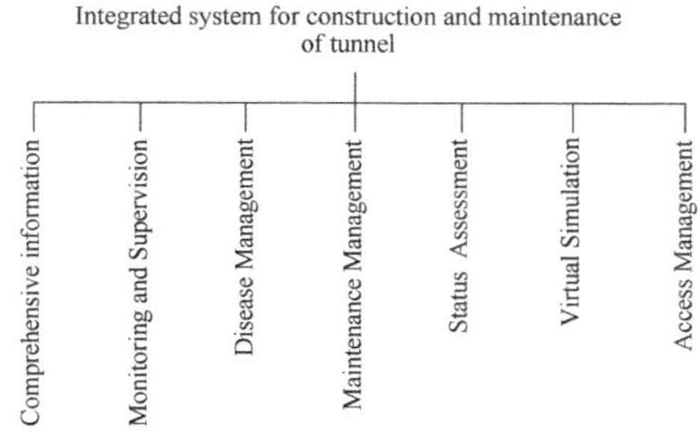

Fig. 7.20 System's functional framework

7.4.3 *Database Technology of Structural Engineering Construction and Maintenance Integration for Tunnels*

Digital tunnel engineering data describes the spatial, temporal, and semantic relationships of various entities and phenomena in tunneling. Due to data being an abstract and approximate description of strata and tunnel structures and their ancillary features based on surveys and measurements, it possesses characteristics in terms of space, time, and scale.

Tunnel engineering data involves geometric, property, topological, and temporal features and corresponding metadata. "Spatial data" refers to information that identifies the geological location, distribution, shapes, or spatial relationship of geometric objects. "Temporal data" maintains temporal topology and represents sequences of all changing data, like engineering construction and monitoring records. "Property data" offers information that is unrelated to spatial locations or temporal topology and is expressed through property items and corresponding property values. Metadata describes geometric and property data.

7.4.3.1 Overall Database Principles

1. Principles of spatial data classification and coding

Spatial data classification and coding should follow the principles of scientificity, systematicity, and practicality.

Scientificity primarily pertains to the rationality of classification and coding, the completeness of the classification system, and the integration of disciplines. When establishing a classification and coding system for tunnel engineering information, comprehensive consideration should be given to the various discipline information involved in the research objectives, and all the information should be incorporated into the same system for current and future overall applications.

Systematicity focuses on the organic and harmonious unity among various parts of the system, emphasizing the general performance of the system, the generalization of commonalities, and the simplicity of the structure.

Practicality implies that data classification should integrate geographic investigation, geological surveys, and subject categories in engineering structure. Data coding should be performed after the classification, and a database storage system and data application model should be established.

Coding, in this context, refers to the process of representing various spatial elements with symbols that are easily recognizable by computers and humans. It is also a technical means of achieving unity in understanding or ideas and of exchanging messages with each other. The direct product of coding is a code. As a foundational task, a reasonable underground spatial information coding system will increase the use of database and enhance the efficiency of information sharing, extending its lifecycle.

2. Principles of tunnel engineering database design

The comprehensive information database for digital tunnel engineering should convert various geological profiles into digital maps for computer storage. To facilitate the analysis and processing of various geological information in secondary development, specific patterns should be used to decompose underground spatial elements. This enables the analysis, presentation, and explanation of different types of entities on the ground or underground, including geological bodies, tunnels, ancillary facilities, underground pipelines, and monitoring facilities. Therefore, the data organization of the comprehensive information database follows specific patterns. The spatial database should encompass the following aspects:

① Seamless organization of underground spatial databases logically and physically.
② High-efficient data access.
③ Easy to maintain database integrity and consistency.
④ Convenient for database expansion.

7.4.3.2 Tunnel Engineering Data Classification, Naming, and Coding

1. Tunnel engineering data classification

The classification of tunnel data focuses on current sources, characteristics, and investigation methods of geoscientific and engineering data. It also takes into account the requirements of creating a 3D model simulation and visualization, virtual viewing, spatial analysis, business applications, and online publishing for databases. This is aimed at providing data support for the management and application services system of tunnel engineering within rail transit projects. Referring to the classification standards of China's land resources, urban geological investigation, and the management and application requirements of digital engineering, data is classified into three levels, covering geography, strata, tunnels, and ancillary buildings (structures), as shown in Table 7.6.

Table 7.6 Classification of tunnel data

First level classification	Secondary classification	Third level classification	Classification code	Remarks
Basic geography (A)	Terrain (A)	Land form (A)	AAA	Contour lines obtained from DEM; other elevation data obtained from DLG, DOM, DRG
		Contour (B)	AAB	
		Other elevation data (C)	AAC	
		Vegetation (D)	AAD	
		Others (0)		
	Surface (B)	Administrative region (A)	ABA	Mainly classified based on topographic map information, important buildings (structures) are expressed as needed
		Water system (B)	ABB	
		Transportation (C)	ABC	
		Geodetic control network (D)	ABD	
		Residential area (E)	ABE	
		Public area (F)	ABF	
		Land use (G)	ABG	
		Scenic area (H)	ABH	
		Important buildings (structures) (I)	ABI	
		Others(0)		
Basic geology (B)	Regional geology (A)	Stratum (A)	BAA	The strata can be divided into two levels; structure can be divided into three levels
		Structure (B)	BAB	
		Rock (C)	BAC	
		Geotechnical (D)	BAD	
		Other(0)		
	Quaternary geology (C)	Sedimentary framework (A)	BCA	According to disciplinary classification, quaternary geology and regional geology are considered as identical, primarily categorized based on sedimentary characteristics
		Tectonic framework	BCB	
		Special stratum (C)	BCC	
		Velocity profiling(D)	BCD	
		Sedimentary characteristics (E)	BCE	
		Other(0)		
	Basic geological exploration (D)	Bedrock geological drilling	BDA	Classification based on basic geological exploration methods
		Cenozoic geological drilling	BDB	
		Exploratory well	BDC	

(continued)

Table 7.6 (continued)

First level classification	Secondary classification	Third level classification	Classification code	Remarks
		Exploratory trench	BDD	
		Exploratory cave (E)	BDE	
		Physical exploration (F)	BDF	
		Chemical exploration (G)	BDG	
		Other(0)		
Engineering geological (C)	Exploration	Drilling (A)	CAA	Classification by engineering geological exploration
		Physical exploration	CAB	
		Chemical exploration	CAC	
		Exploratory well	CAD	
		Exploratory trench	CAE	
		Vertical shaft	CAF	
		Audit	CAG	
		Others		
Engineering geology (C)	In situ testing (B)	Vane test (A)	CBA	Classification based on in situ testing content
		Standard penetration test (B)	CBB	
		Pressure meter test (C)	CBC	
		Load test (D)	CBD	
		On-site direct shear test (E)	CBE	
		Wave velocity test (F)	CBF	
		Dynamic penetration test (G)	CBG	
		Static penetration test (H)	CBH	
		Resistivity logging (D)	CBI	
		Water pumping test (J)	CBJ	
		Water injection test (K)	CBK	
		Flat dilatometer test (L)	CBL	

(continued)

Table 7.6 (continued)

First level classification	Secondary classification	Third level classification	Classification code	Remarks
		In situ stress testing (M)	CBM	
		Excitation method testing (N)	CBN	
		Others (0)		
	Indoor test (C)	Physical and mechanical tests (A)	CCA	Classification based on indoor experimental content
		Compression consolidation test (B)	CCB	
		Shear test (C)	CCC	
		Dynamic test (D)	CCD	
		Rock testing (E)	CCE	
		Others (0)		
	Achievement data (D)	Basic information of strata (A)	CDA	Classification based on the types of processing of raw data
		Plane distribution of drilling holes (B)	CDB	
		Individual drilling information (C)	CDC	
		Thread hole data (D)	CDD	
		Physical and mechanical parameters of the stratum (E)	CDE	
		Comprehensive results of other data sources (F)	CDF	
		Others (0)		
Hydrogeology (D)	Exploration (A)	Hydrological drill (A)	DAA	Classification by hydrogeological exploration method
		Shallow shaft (B)	DAB	
		Observation well (C)	DAC	
		Others (0)		
	In situ test (B)	In situ pressure test (A)	DBA	Classification based on in situ testing content
		In situ water injection test (B)	DBB	

(continued)

Table 7.6 (continued)

First level classification	Secondary classification	Third level classification	Classification code	Remarks
		In situ pumping test (C)	DBC	
		In situ groundwater permeability test (D)	DBD	
		Others (0)		
	Indoor testing (C)	Physical and mechanical tests (A)	DCA	Classification based on indoor experimental content
		Compression consolidation test (B)	DCB	
	Achievement data (D)	Hydrological drill layout table (A)	DDA	Classification based on the types of processing of raw data
		Observation well layout table (B)	DDB	
		Table of observation data for confined water level (C)	DDC	
		Evaluation table for liquefaction of foundation soil (D)	DDD	
		Water quality analysis table (E)	DDE	
		Others (0)		
Environmental geology (E)	Regional environmental geology (A)	Desertification (A)	EAA	Mainly classified from the aspects of regional environment, water pollution, environmental pollution, and geological disasters in environmental geology
		Salt intrusion zone (B)	EAB	
		Others (0)		
	Water pollution (B)	Groundwater pollution (A)	EBA	
		River and lake pollution (B)	EBB	
		Others (0)		
	Environmental pollution (C)	Soil pollution area (A)	ECA	
		Endemic disease distribution (B)	ECB	
		Mine pollution (C)	ECC	
		Garbage stacking distribution (D)	ECD	
		Others (0)		

(continued)

Table 7.6 (continued)

First level classification	Secondary classification	Third level classification	Classification code	Remarks
Environmental geology (E)	Geological disaster (D)	Land subsidence distribution (A)	EDA	Mainly classified from the aspects of regional environment, water pollution, environmental pollution, and geological disasters in environmental geology
		Sea level rise and tidal disasters (B)	EDB	
		Slope instability (C)	EDC	
		Subsoil deformation (D)	EDD	
		Sand liquefaction (E)	EDE	
		Methane in shallow stratum (F)	EDF	
		Soil and water pollution (G)	EDG	
		Land subsidence (H)	EDH	
		Soil salination (D)	EDI	
		Groundwater waterlogging (J)	EDJ	
		Others (0)		
Underground structure (F)	Underground pipelines (A)	Water supply (A)	FAA	For the sake of clarity in classification, based on the second level classification of underground structures, the third digit is used as the classification code for digital engineering, represented by a single digit: for example, the engineering design information is 1; the engineering construction information is 2; the engineering monitoring information is 3; the engineering inspection information is 4. Other engineering information can be inferred in this way, and there is no need to set this information code to "0"
		Drainage (B)	FAB	
		Gas (C)	FAC	
		Thermodynamics (D)	FAD	
		Industrial (E)	FAE	
		Electricity (F)	FAF	
		Telecommunication (G)	FAG	
		Comprehensive (H)	FAH	
		Others (0)		
	Tunnel (F)	Main body (A)	FFA	
		Road surface (B)	FFB	
		Wall (C)	FFC	
		Approach (D)	FFD	
		Shaft (E)	FFE	
		Fan (F)	FFF	
		Luminaire (G)	FFG	
		Working shaft	FFH	
		Pump station	FFI	

(continued)

Table 7.6 (continued)

First level classification	Secondary classification	Third level classification	Classification code	Remarks
		Others (0)		
Model data (G)	Excavation evaluation model (A)			
	Obstacle analysis model (B)			
	Underground space planning model (C)			
Others (0)				

2. Database naming and coding content

In the establishment of tunnel engineering databases and 3D models, uniform coding and naming are required. This report refers to relevant standards and takes into account the characteristics of tunnel engineering management and application systems to formulate unified naming and coding rules for tunnel engineering database construction. The rules are made for layer naming, database and data sheets naming, entity coding naming, index coding naming, and metadata table naming.

(1) Layer naming

Layers are basic management units for graphical entity data based on geographic, geological, and structural aspects. Layers are divided into points, lines, polygons, and volumes. To ensure the independence of graphical information and related property information and prevent duplicate layer names, the layer naming structure is as follows:

△△△△△△△△△X X X X
11,111—⑤ Element property code and layer number
III1—④ Frame code
II1—③ Sublayer code
I1—② Thematic layer code
1—① Project code.

(2) Database naming

Data is stored in the same physical database or in a heterogeneous or distributed database. The naming and coding structure for databases is as follows:

△△△△△△X X
I I 1—③ Database numeric code
I 1—② Data category code
1—① Project code.

(3) Entity naming

Entity coding is set up for the management and querying of entity spatial data and property data. It treats the layer as a whole, with features as the basic units (entities). In the same layer, each entity has a unique ID value and corresponds to the respective property data record. The coding principles are as follows:

△ X X X X X
I 1—② Entity code
1—① Entity classification code.

(4) Data sheets naming

A data sheet is a subset of the database or the primary table for graphical properties. Layer property tables should correspond one-to-one with layers, and their naming rules are the same as layer naming. For one-to-many sub-tables, they can be associated through key fields, and their naming rules follow the data sheet naming rules.

Data sheets and graphical property data sub-tables under database management are named and encoded as follows:

△ △ △ △ △ XXX
I I I 1—④ Data sheets number
I I 1—③ Table type code
I 1—② Detail level code
1—① Data classification code.

(5) Index naming

Data index is primarily used to identify data records in data tables for point spatial databases. The coding principles are as follows:

△ X X X X X X
—③ Data record number
—② Discipline code
—① Data source code.

(6) Metadata table naming

Metadata is data about data, describing information about data and information resources. Metadata is mainly used for information sharing. Users can use database to search for data collections from which data volume, data accuracy, owner information and other information can be seen. And they can also find other resources, including data format, spatial coordinate systems, data representation, software and hardware environment for data use, data usage guidelines, and data standards. Therefore, in the process of data integration and exchanges, metadata can help realize spatial data matching, property consistency processing, data conversion between platforms, and more. Metadata table naming is as follows:

I1 ② Table name
1—① Metadata identifier.

(7) Field naming

According to reference standards, for convenience, field names in data tables, except for some geometric information, are all named in Chinese.

7.4.3.3 Tunnel Engineering Database Design

1. Logical design

During the logical design, two main tasks need to be completed: the design of the picture frame structure and the organization of frame information. The comprehensive underground space database integrates both logical aspects.

(1) Design of frame structure

The access methods for the underground space comprehensive information database include centralized storage and sectional storage, which one you choose depends on the comparison and contrast of application environments. Generally, the primary factor is data volume, with a control value usually set at 1 GB. If the data volume is less than 1 GB, centralized storage is suggested; otherwise, sectional storage is used.

Sectional storage inevitably leads to situations where a spatial entity crosses different picture frames, meaning that the spatial entity is divided into spatial basic units. This presents challenges for the integrity and consistency of entities. Therefore, it is necessary to establish connection between picture frames, spatial entities, and spatial basic units using a spatial index table.

(2) Design of layer structure

Certain related graphic elements are grouped together to form layers based on specific user needs or standards. All data within a layer shares the same geometric characteristics and similar property characteristics. Layers can be organized based on spatial feature categories or thematic divisions. System information is divided into thematic layers, with each thematic layer further divided into several sub-layers.

2. Physical design

The physical design primarily is to implement the logical structure of the spatial database on actual physical storage devices, involving the design of the database's physical structure and the physical construction of the database.

The comprehensive information database for digital tunnel engineering adopts a seamless data storage model hierarchy, as shown in Fig. 7.21. The digital tunnel serves as the root of the tree, representing a complete comprehensive tunnel information database. Ground objects in the underground space must maintain consistency and integrity in terms of storage and representation. According to the principle of picture frame division, spatial data is divided into multiple picture frames, with each picture frame forming a node in the tree. Based on the hierarchical principle mentioned in the previous section, picture frames are divided into multiple layers. Each type of feature in a layer is defined by different files (data sheets), forming the leaf nodes of the tree. This structure creates an internal spatial indexing system.

Fig. 7.21 Physical structure of underground space comprehensive information database

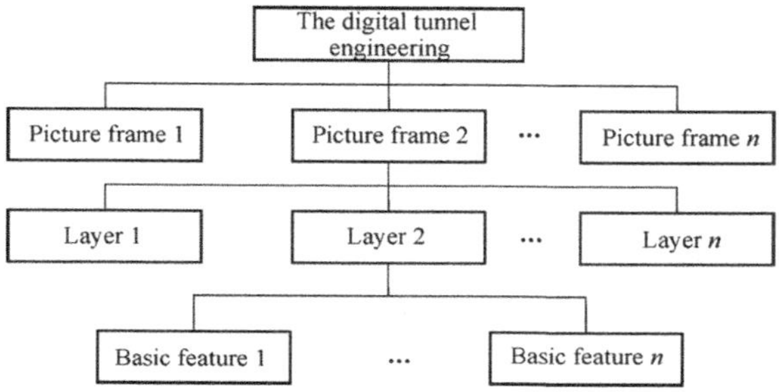

3. Organization and management of spatiotemporal data

(1) Organization of spatiotemporal data

The organization of spatiotemporal data first emphasizes the temporal cohesion and spatial and property cohesion of the data. Given strong cohesion, where objects undergo frequent changes, significant changes, or frequent operations on historical data, such as construction monitoring data, current data should be stored together with historical data in the same data set. Given weaker cohesion, where current data is frequently accessed and historical data is less manipulated, it is advisable to consider separating current and historical data storage. If there is strong cohesion between spatial and property data, then in database design, spatial and property data should be closely associated and stored in a single data set; otherwise, they should be stored separately.

(2) Management and representation of spatiotemporal data

To facilitate the management and representation of spatiotemporal data, the concept of time vectors is introduced. All changes are stored as time series, explicitly capturing temporal changes and temporal topological relationships. Through such a timeline or time vector, spatiotemporal data can be managed and operated simply and directly.

In temporal databases, time stamp is commonly used to manage temporal property data. Two commonly used methods are marking time stamps to tuples and to property items. The former involves adding time stamps to property tuples for the same entity, while the latter involves time marking each property item.

The tuple time stamping approach uses several tuples with time stamps to describe the excavation progress of a specific section, which intuitively resembles a construction record table. However, for the database, this approach results in redundant data by repeatedly recording unchanged fields. Time stamping by property item actually transforms construction historical data expressed in the first normal form into non-first normal form relations.

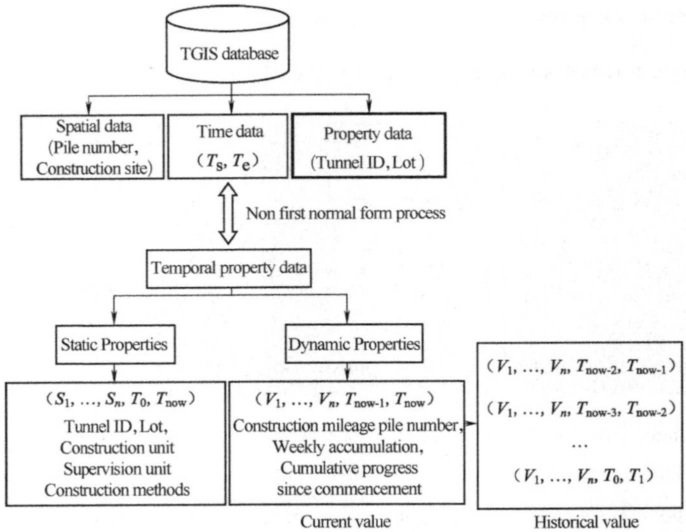

Fig. 7.22 Organizational structure of temporal attribute data

Compared to simple time data, time stamping property data while linking them with entity identifiers better reflects the characteristics of a temporal database structure. The organization and management of spatiotemporal data are illustrated in Fig. 7.22.

7.4.4 Three-Dimensional Data Modeling and Visualization Techniques for Tunnels

Three-dimensional models, as tools for information retrieval and carriers of information representation, provide 3D results and visualizations for tunnel basic information, structural condition evaluations, and disease information. Through virtual exploration, users can view the surrounding environment, geological layers, design information, structural disease, maintenance data, long-term settlement, cross-sectional convergence, tunnel conditions, etc. The system integrates various types of information on the interface, making it easier for professionals to conduct comprehensive analysis.

Within the platform, the VR technology utilizes well-established tools like MultigenCreator and VegaPrimer as the development kit. MultigenCreator is a powerful and interactive tool for 3D modelings which can be used for large-scale visual simulation. VegaPrimer is employed for real-time visual analog, virtual reality, and general visual applications, offering convenient creation, editing, and driving tools, providing a complete application interface, and maximizing software control and flexibility for software developers to develop specific functions as needed.

In this platform, parameterized modeling is based on the OpenFlightAPI for shield tunnel procedural modeling. OpenFlightAPI serves as a secondary development tool for MultigenCreator, a professional modeling software. For each type of engineering structure, including structural components, disease information and other entity objects, if there are existing component or cross-sectional models, OpenFlightAPI is used to assemble them following a certain organizational relationship, based on the positioning information from the database. The result is to generate Fit model files that meet the requirements of the VP program.

7.4.4.1 Three-Dimensional Geological Modeling Technology

Tunnel design, construction, and operational maintenance typically focus on the spatial distribution of geological layers, and therefore, tunnel digitalization aims to establish geometric models of the layers. In this context, the layer modeling is underpinned by the triangular prism data model. This method relies entirely on drilling data without human intervention, particularly suitable for handling layered geological situations, as well as addressing geological problems such as strata folding, pinch-out, and crystal penetration.

1. Geological modeling design based on drilling information

The modeling process is divided into three steps: The first step is to establish geological interfaces that can describe geological conditions, such as lens, immersed bodies, and pinch outs, based on drilling data. Meanwhile, in the early stage of modeling, data pretreatment will help form the pinch-out profiles of each interface; The second step is to build a stratigraphic bounding box for the research area by the profiles in the first step; The third step is to create a fault plane based on the profile information, insert it into the hermetic solid stratigraphic model that is previously established and filled with interface information and then move the internal model interface based on the fault dislocation information, as shown in Fig. 7.23.

2. Simple layered earth modeling method

Owing to the layered distribution of the strata, a standard sequence of strata can be developed based on the stratigraphic divisions in all drillings. If a drilling loses some of its strata, a virtual zero thickness stratum is inserted into the drilling according to the standard sequence. After such pretreatment, all drillings will have the same stratigraphic divisions.

Based on the x, y coordinates of all drillings, a planar Delaunay triangulation algorithm is used to establish a surface triangle mesh. As this mesh serves as the reference plane for 3D stratigraphy, it is called surface reference mesh.

Because of the identical layer sequences in all drillings, the entire 3D stratigraphy can be constructed by starting with a surface reference mesh and extending the triangular sheets from top to bottom into prisms. During the generation of prisms, if

Fig. 7.23 Process of three-dimensional geological modeling

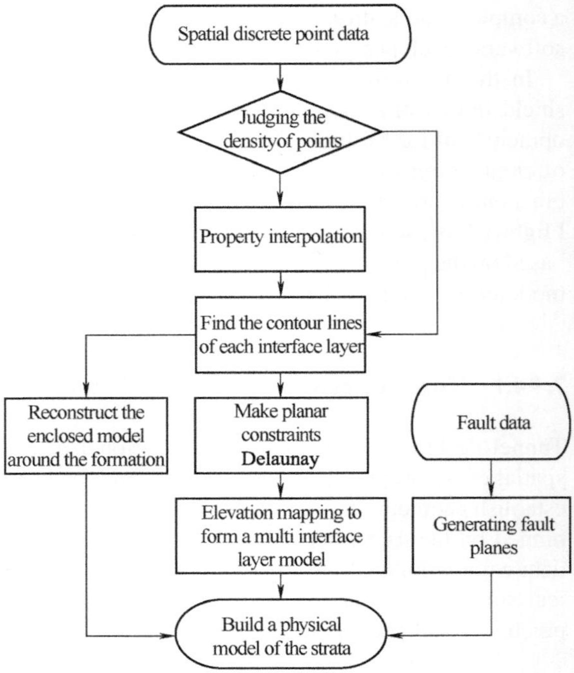

all three layers in a prism have zero thickness (representing virtual layers from the drillings), then that prism does not need to be generated. If one or two layers have zero thickness, prism degradation is applied.

3. Modeling method for interlayered geological layers

To accurately reflect the stratigraphic distribution of the tunnel construction site, this study proposes a modeling method that includes interlayered geological layers which is based on the simple layered geological modeling approach. First, the interlayered formations are pre-identified and modeled separately. Then, the interlayer information is removed from the drillings. After modeling the geological layers using the simple modeling method, the interlayer models are inserted into their respective geological layers.

4. Modeling process details

First, identify the drillings that contain interlayered formations. Then, independently model the interlayered formations and remove the interlayer information from the drillings once modeling is complete. Next, use the stratigraphic modeling methods to build the geological layer model. Finally, insert the interlayered formation models into the geological layer model. Additionally, if it is possible to independently model interlayered formations within multi-layered geological formations, the modeling method for simple layered formations can also be extended to such conditions by "merging" the multi-layered formations into single-layer formations. This allows the simple layered geological modeling method to be applied more broadly in such scenarios.

7.4.4.2 A 3D Technology for Tunnel Automated Modeling

1. Design Philosophy

Automated 3D modeling is based on parametric programs instead of human machine interactions. Real-time modeling is achieved through network service programs that provide timely and responsive updates.

Additionally, considering the efficiency during system operation, to avoid loading too many model files at once, which could lead to low system efficiency, level of detail (LOD) is used to create hierarchical models, facilitating large-scale terrain management and thereby improving system performance.

2. Modeling principles

Based on the summarization of actual underground engineering objects, most models, such as tunnel segments, synchronized construction components, and structural disease, can initially be located on cross sections. Subsequently, they can be further positioned by ID, route, mileage, and rotation information. For certain special model objects, supplementary information can also be included to fulfill specific organizational relationships, such as time and type information. Consequently, most models can be processed as the same type, and by configuring the organization through a configuration file, the main and supplementary information can be retrieved from the database.

Regarding the modeling approach, models can be categorized into two methods: parts assembling and section stretching. For instance, modeling precast tunnel segments involves the first method, while components like corbels are achieved through the second method.

From an organizational perspective, modeling can be categorized into temporal and non-temporal organizational relationships. For example, in models related to tunnel segments, both design and construction models are organized by temporal relationships, allowing for functions such as time-based construction progress playback.

Within a specific model category, internal classifications can be made based on certain rules. For instance, in the context of disease information, water leakage can be differentiated based on parameters like area, length, and width. Different-sized component models can then be invoked to represent the various types of disease.

In summary, a configuration file can be used to generalize the content of most models within a specific category. This file includes ID, information on route, mileage, rotation, and time, modeling approach and type rules. Among these, ID, information on route, mileage, and rotation, and modeling approach are necessary information, while time information and type rules are supplementary. If left

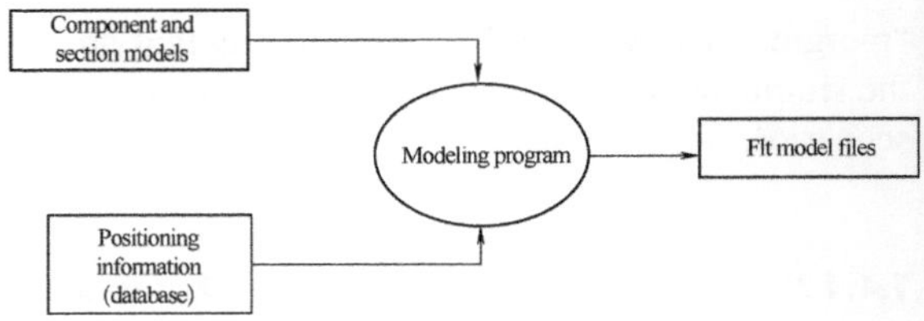

Fig. 7.24 Basic design ideas for three-dimensional structure modeling

blank, they are processed according to default rules. The three-dimensional structural modeling approach is depicted in Fig. 7.24.

3. Modeling process

The three-dimensional modeling process, as shown in Fig. 7.25, is divided into six main steps:

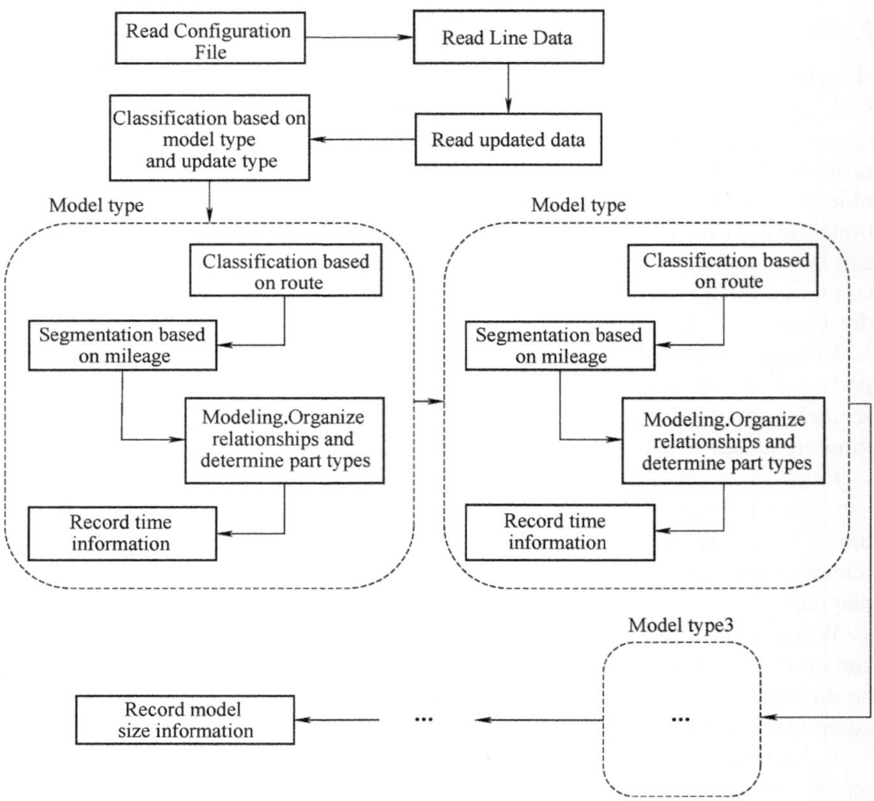

Fig. 7.25 Structural modeling process

(1) Read configuration files: For each category of objects that need to be modeled, the required modeling information corresponding to database field names is stored in a configuration file, through which, the automated modeling program will read the relevant data. Such files also contain information for determining component types to decide which component model to call during modeling.
(2) Read route data: Read data from the "Route Design Table", "Plane Axis Tables", and "Vertical Axis Tables" in the database to obtain route data for axis algorithms. Additionally, retrieve the information from axis algorithms for organizing and categorizing model segments.
(3) Read update data: Read data from the "Data Update Table" in the database, extract the data, and clear the records of data that was read during this session.
(4) Categorize retrieved data: First, categorize the data retrieved based on the model's corresponding data table name in the configuration file. Then, classify the data for each model category based on update type and route classification.
(5) For each model category, after categorizing the update data by update type and route classification, determine which segment of the FLT model file needs to be modified based on mileage and segment length. Perform the necessary operations on that segment model. For instance, if a model needs to be added, call the appropriate component model based on the evaluation rules in the configuration file and add it to the model file at the specified location. If the model category includes time information, organize the model files based on time relationships and record the starting time information for that segment model in the corresponding XML file.
(6) Repeat step 5 for each model category related to the update data and record the filenames of all modified model files.

Based on the filenames recorded in step 6, traverse the model files and record the dimensional information of the models for the large-scale terrain management in the VP programs.

7.5 Summary

In response to the intelligent operation and maintenance management requirements throughout the entire lifecycle of tunnels, a comprehensive lifecycle system of intelligent operation and maintenance management for tunnels has been developed, utilizing technologies such as the Internet, the Internet of Things (IoT), geographic information systems (GIS), and 3D digitization. This system includes the Geological Information Management and Analysis System, Supervisory Measurement Data Management System, Ground Inspection Management System, and Construction and Maintenance Data Integration System, enabling the integrated management of massive data and multi-source information. It has successfully achieved full-coverage and efficient operation and maintenance management.

1. "Geological Information Management and Analysis System for Tunnels": As tunnel structural deformation is highly correlated with geological conditions, the system integrates the characteristics of over 12,000 drillings and the ground settlement information over last 20 years, allowing for the stratigraphic visualization along tunnel routes, intelligent geological risk alerts, and smart emergency response recommendations. The system's achievements have been widely applied in multiple cities in the Yangtze River Delta.
2. "Monitoring and Measurement Data Management System for Tunnels": Designed to address challenges in extracting, storing, cleaning, and utilizing a vast amount of heterogeneous data from structural monitoring measurements in soft soil tunnels, the system, based on China's "Tianditu" platform, pioneered a monitoring and management system for tunnel in China. It achieves the integrated management of multi-source data from "air + ground + underground" remote sensing images, laser scanners, sensors, and precision engineering measurements, enabling comprehensive and efficient control over the impact of nearby construction projects on subway structures. The system was awarded grand prize in the "Second Tianditu Application Development Competition (2014)" in the surveying and mapping industry. It has been widely adopted by subway enterprises in Chinese cities, covering more than 1000 projects and over 200 million data sets.
3. "Ground Inspection Management System for Tunnels": To solve low inspection efficiency in the protection zones of large-scale tunnels networks and the difficulty of monitoring unauthorized construction activities, the system established a big data platform for managing inspection trajectories and unauthorized projects, enabling "high-efficiency, full-coverage" inspections in subway constructions. It has been deployed in Shanghai, Hangzhou, and other regions, with over 30,000 tasks executed, more than 20,000 inspections conducted, and nearly 50 million trajectory data points collected. It has successfully identified over 3000 unauthorized projects, thereby avoiding safety hazards of tunnel breaches.
4. "Data Integration System Based on Construction and Maintenance Integration": This system has tackled challenges in the integration of multi-source and mass data sets. To address issues arising from the large data for tunnel operation and maintenance, data disunity, and information silos, it leverages the VR technology for tunnel disease treatment based on 3D data modeling. This integration system fuses basic data throughout the entire lifecycle of tunnel structures, reaching higher informatization and intelligence in subway tunnel operation and maintenance management.

Chapter 8
Engineering Applications

The research findings discussed earlier encompass a range of integrated technologies, equipment, and systems designed to ensure the tunnel safety of metro operating in soft soil areas. They include reinforcement-integrated technologies and equipment such as micro-disturbed grouting and internal intension steel rings used when repairing large deformation and severe damage of tunnels. More are included, like rapid comprehensive detection mobile equipment to assess the safety of metro tunnel structures, data management systems for monitoring and measuring rail transit, and intelligent operation and maintenance management systems, etc. These technologies, equipments, and systems provide real-time perception, forecast and early warning of security risks and timely assessment of the security, allowing for effective control or elimination of security hazards in tunnel structures. By systematically addressing structural security, repair and maintenance challenges, they effectively avoid prolonged shutdowns and repairs of metro tunnels, minimizing the social impacts. Additionally, they reduce the cost of major repairs for operating metro tunnels and provides strong technical support for prolonging the durability of metro tunnels.

The related research achievements have been widely applied in dozens of metro tunnels in more than 10 cities, including Shanghai, Guangzhou, Nanjing, Wuhan, Kunming, Hangzhou, Ningbo, Shenzhen, and Suzhou. They have been applied in over 100 metro tunnel projects. This chapter focuses on 13 representative engineering project cases, as shown in Table 8.1.

R. Wang et al., *Structural Safety Early Warning and Treatment Technology of Metro Tunnel in Soft Soils*, Key Technologies for Tunnel Construction under Complex Geological and Environmental Conditions,
https://doi.org/10.1007/978-981-96-6410-8_8

Table 8.1 Table of engineering application in this chapter

Number	Technical category	Engineering cases
1	Micro-disturbed grouting	Convergence treatment of Shanghai Metro Line 10 between "Yingao Road Station" and "Xinjiangwancheng Station"
2		Tunnel deformation treatment of Nanjing Metro Line 2 between "Yurundajie Station" and "Yuantong Station"
3		Micro-disturbed grouting of Shanghai Metro Line 11 for a crossing area of Beiheng passageway
4	Internal intension steel ring	Tunnel structure repair of Shanghai Metro Line 2 between "Chuangxin Middle Road Station" and "Huaxia East Road Station"
5		A tunnel structure repair of Shanghai Metro Line 7 between "Changqing Road Station" and "Houtan Road Station"
6		Tunnel repair of Nanjing Metro Line 2 between "Jinma Road Station" and "Maqun Road Station"
7		Tunnel repair of Wuhan Metro Line 2 Phase One between "Changgang Road Station" and "Hankou Railway Station"
8	3D laser scanning detection	Annual convergence detection of mobile laser scans of Shanghai Metro Line 2, Line 10, and Line 12
9		Shield tunnel convergence detection of mobile laser scans of all Nanjing Metro Lines
10		Tunnel structure verification based on laser scanning (take the metro tunnel structure detection in operation of "China-US Trust" site as an example)
11	System platform	Application of the structure monitoring data management platform of Shanghai Rail Transit
12		Application of the patrol platform of Dongguan Metro Protection Zone
13		Application of remote monitoring for foundation pit metro structure of "Hengda Center" in Shenzhen

8.1 Engineering Cases of Micro-disturbed Grouting

8.1.1 A Convergence Treatment of Shanghai Metro Line 10 Between "Yingao Road Station" and "Xinjiangwancheng Station"

8.1.1.1 Background of the Project

Shanghai Metro Line 10 is designed in a "Y" shape, starting from "Xinjiangwancheng Station" and ending at "Hongqiao Railway Station" and "Hangzhong Road Station". It branches at "Longxi Road Station" and the entire line is located underground. Line 10 passes through multiple transportation hubs and commercial districts, serving as a crucial route across the city. The geological conditions of the section in this case are challenging, with a deep layer of silty sand and dust of ②$_3$. During the shield tunnel construction, controlling the shield posture becomes difficult, resulting in issues such as staggered joints, missing pieces and significant convergence after segment assembly. In 2009, affected by the excavation of an adjacent foundation pit, the cumulative maximum convergence of the section exceeded 100 mm, resulting in a higher risk for the tunnel safety. Prior to that, there was no precedent for dealing with tunnel convergence of shield tunnels in soft soil at home and abroad. Following expert consultations, a newly developed micro-disturbed grouting technique was chosen for the convergence treatment of this section, marking the first such treatment in Shanghai's soft soil conditions. The first grouting treatment was conducted in mid-November, 2009, and the second grouting remediation in March, 2011.

8.1.1.2 Construction Process

In 2009, the first-time grouting was arranged parallel to the tunnel axis within the range of 315–355 rings on the plan, with a total of three rows. The center-to-center spacing between the piles in the first row and the tunnel axis was 4.3 m.

In 2011, the second-time grouting was conducted approximately 50 m outside the upward section of the tunnel from "Xinjiangwancheng Station" to "Yingaodong East Road Station" (within the range of 315–355 rings), with 41 holes set up for grouting.

The grouting holes were arranged in a plum blossom pattern, spaced 1.2 m apart horizontally and vertically. The layout of the grouting holes is shown in Fig. 8.1. The grouting depth for each hole corresponds to the depth of the tunnel, which is 6.2 m. The grouting progresses from the inside out, and within the same row, a drilling-one-and- skipping-two construction sequence is followed.

During the first emergency repair period (November 16, 2009 to December 11, 2009), double liquid grouting reinforcement was completed within the respective ranges of 4.8 and 3.6 m. During the second reinforcement period (February 21, 2011 to April 10, 2011), double liquid grouting reinforcement was completed within the

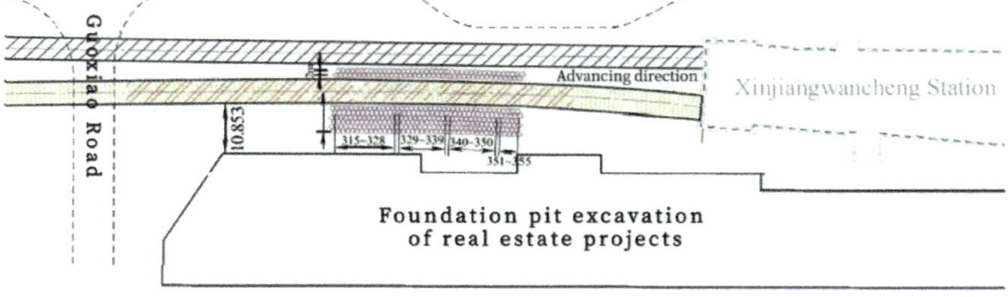

Fig. 8.1 Schematic layout of a grouting hole (unit: mm)

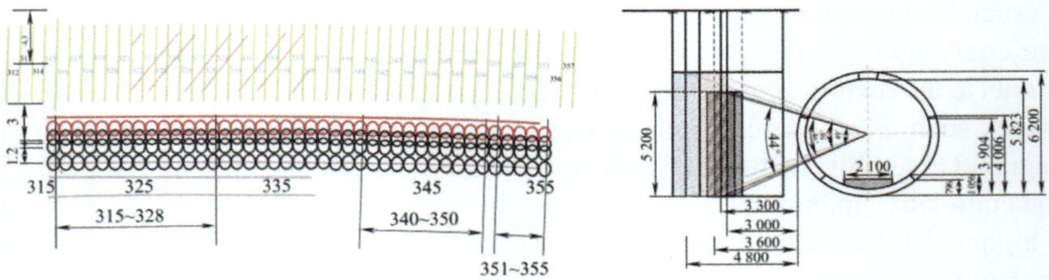

(a) Plan view (b) Cross-section view

Fig. 8.2 Micro-disturbed grouting

respective ranges of 3.0 and 3.3 m. The schematic illustrations of micro-disturbed grouting plan and cross section are shown in Fig. 8.2.

8.1.1.3 Construction Effect

The construction began on November 16, 2009. After grouting of 3.6 and 4.8 m rows, the convergence of the tunnel decreased to 2.7 to 26 mm on December 11, 2009. Subsequent measurements showed further reductions, with convergence ranging from − 4.97 to − 14.78 mm on February 21, 2011, 7.43 to 29.78 mm on April 10, 2011, and 1.1 to 27.9 mm on February 27, 2012.

8.1.1.4 Analysis and Conclusion

This project marked the first implementation of micro-disturbed grouting in an engineering practice. The construction verification of the micro-disturbed grouting technique in this project demonstrated a significant effect in controlling the convergence of the tunnel.

The construction verification also confirmed the practical effectiveness of the dedicated flow meter, mixer and grouting nozzle developed for this method. It proved that the flow meter with a display screen played an important role in real-time monitoring and controlling the grouting flow rate and pressure. Additionally, its data storage function provided strong support for subsequent grouting data recording and analysis.

The specially designed mixer thoroughly blends cement and water glass, before grouting into the soil, ensuring the quick solidification of the slurry and guaranteeing the grouting effectiveness and quality. The dedicated grouting protection device invented in this method has demonstrated excellent performance during construction and can be reused multiple times. It was observed that the grouting nozzle was prone to premature opening and blockage with soil during the lowering process. Therefore,

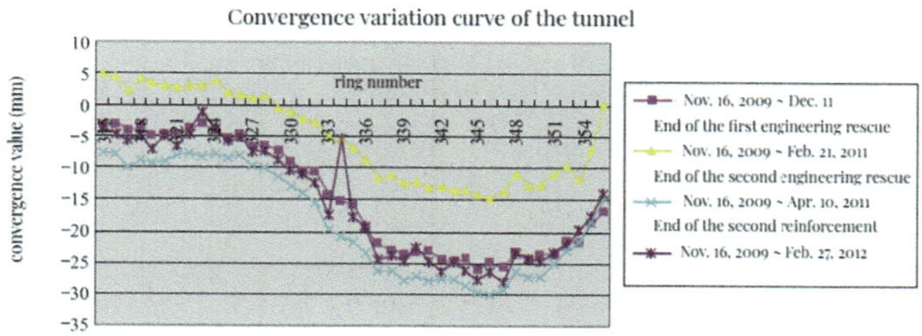

Fig. 8.3 Effectiveness of two grouting reinforcement treatments

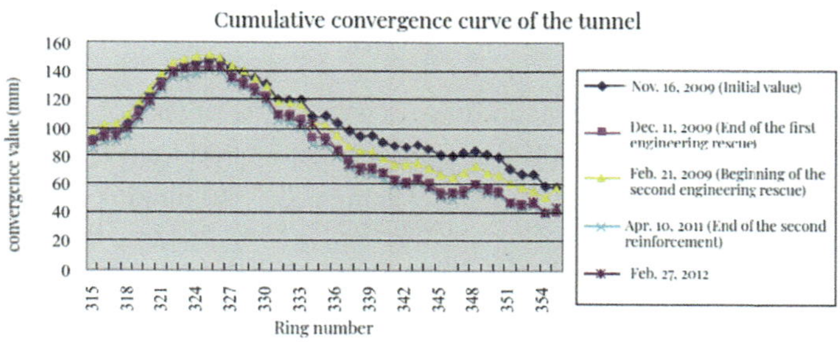

Fig. 8.4 Cumulative convergence of the tunnel

it is recommended to wrap the grouting nozzle with insulating cloth before lowering, as shown in Figs. 8.3 and 8.4.

In addition, this project also assessed the effectiveness of the micro-disturbed grouting technique under various parameters. This included different distances between tunnels, various starting and ending grouting heights, diverse slurry ratios and water-cement ratios, as well as varying grouting volumes. It provided valuable experience and solid foundation for future optimization of construction parameters and scheme development.

8.1.2 A Tunnel Convergence Treatment of Nanjing Metro Line 2 Between "Yurundajie Station" and "Yuantong Station"

8.1.2.1 Background of the Project

The foundation pit for the Nanjing Financial City project is located in the section between "Yurun Street Station" and "Yuantong Station" on Nanjing Metro Line 2. The outer edge of the pit's retaining structure is about 15.0 m away from the nearest section of the metro line, with the pit bottom being 2.15–4.25 m lower than the tunnel crown. Due to the impact of the nearby excavation, a significant convergence occurred in the corresponding section of the tunnel between "Yurun Street Station" and "Yuantong Station" of Metro Line 2. The maximum convergence of the up line and down line reached 93.7 mm, and 86.1 mm, respectively. As the shield tunnel in this section was assembly with staggered joints, such convergence approached the threshold. Numerous longitudinal cracks appeared at the top segments. After tunnel safety assessment, it was determined that micro-disturbed grouting was needed for all tunnel segments with convergence exceeding 70 mm. Specifically, for the down line (left track) tunnel, grouting was required for rings of Number X523-X665, and while for the up line (right track) tunnel, grouting was required for rings of Number S484-S495, S518-S630, and S696-S712.

8.1.2.2 Construction Process

The remediation of this project started on December 9, 2013, and was completed in the early morning of September 26, 2014, lasting nearly 10 months. A total of 1275 grouting holes were completed. The construction was divided into three stages. The first and second stages involved dual-sided and dual-row convergence treatment in two areas with significant convergence, following the original plan. In the third stage, tracking grouting was conducted simultaneously with the pit construction. Multiple rows and multiple grouting operations were performed in the areas where tunnel convergence still exceeded 70 mm. The arrangement of grouting holes is shown in Fig. 8.5.

The tunnel in the ring S695–S714 section of the up line (right track) tunnel is located in a layer of fine sand. This specific section suffered from severe convergence. Difficulties arose after an experimental construction in this area, including challenges in lowering pipes, producing slurry, and no effectiveness of treatment. Through multiple trials, improvements were made to the grouting parameters and construction techniques. These included increasing the grouting volume, performing multiple rows and multiple grouting operations, and strictly limiting the drilling depth of the drilling machine (to reduce disturbance to the surrounding soil). These measures ultimately achieved the goal of convergence treatment.

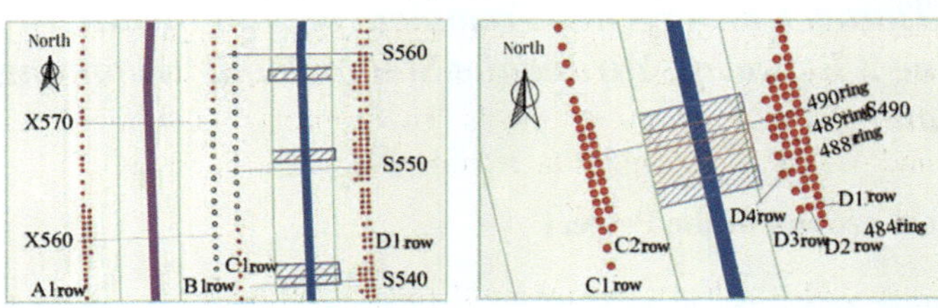

Fig. 8.5 Schematic layout of grouting holes

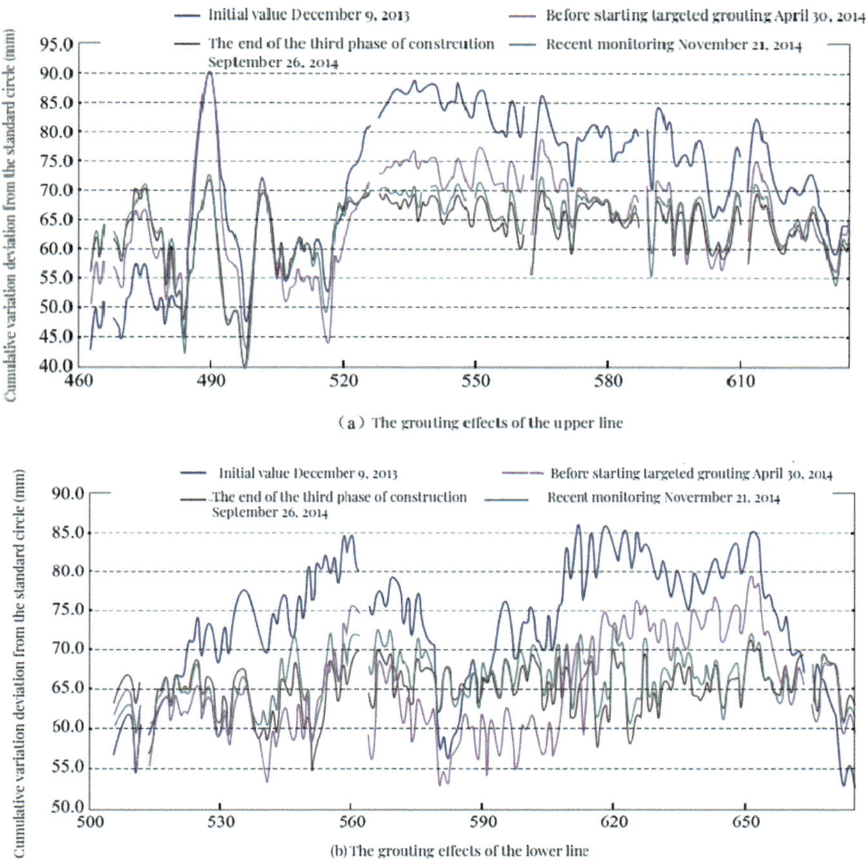

Fig. 8.6 Convergence variation curve of Nanjing Metro Line 2 between "Yurun Street Station" and "Yuantong Station" during micro-disturbed grouting

8.1.2.3 Construction Effect

After finishing the three stages of remediation in this project, the convergence of all the shield tunnels in the construction areas have been reduced to below 70 mm. The convergence problem has been effectively controlled. The variation curve of the tunnel convergence during the construction period is shown in Fig. 8.6.

8.1.2.4 Analysis and Conclusion

The following conclusions can be drawn based on the analysis of the monitoring data during the construction period.

(1) The convergences in both the up line and down line tunnels during the grouting process were more pronounced and followed a consistent trend, characterized by lateral compression and a decrease in tunnel convergence. Up to November 21, 2014, the maximum cumulative variation of tunnel convergence in the up line tunnel was − 20.4 mm (S537), and in the down line tunnel was − 22.1 mm (X611).
(2) During the grouting, the ground around both the up line and down line tunnels exhibited slow and continuous expansion. The expansion is uneven due to the different locations and geological conditions of the two tunnels. Additionally, the overall treatment effect of the two tunnels shows regional differences as a result of the longer construction cycles and variations in the effectiveness of each grouting treatment operation.
(3) After the grouting, the maximum cumulative variation of the convergence of the up line tunnel compared to the standard circle ~~is~~ 72.1 mm (S489), with a cumulative variation of − 16.0 mm during the grouting. The maximum cumulative variation of the convergence of the down line tunnel compared to the standard circle is 73.3 mm (X620), with a cumulative variation of − 9.7 mm during the grouting.
(4) Based on the cumulative variation curve and the duration curve of characteristic points, it can be observed that after the first stage of ground grouting (June 28, 2014), there is a certain amount of recovery in the tunnel diameter in the lateral direction. This recovery also occurs after the second and third stages. However, the magnitude of recovery is relatively small compared to the cumulative variation during grouting, and the tunnel itself exhibits slow convergence, primarily characterized by lateral stretching and an increase in the horizontal diameter.
(5) During the grouting, the initial implementation according to the construction plan led to relatively consistent convergence effects in the tunnel. However, as targeted grouting began, repeated grouting was conducted in areas where the cumulative variation of the tunnel convergence compared to the standard circle was higher. As a result, the majority of the tunnel areas controlled the cumulative convergence within 70 mm.
(6) This project marked the first application of micro-disturbed grouting in a fine sand layer, presenting challenges not encountered in soft soil layers. First, there were difficulties in lowering the pipes, as the existing construction equipment was inefficient in the sand layer, necessitating the development of integrated equipment that combined drilling, vibration, and grouting. Secondly, the pipes were prone to blockage. Moreover, the relatively large gaps in the thread forms of the grouting core pipes allowed the fine sand to easily infiltrate the pipes during lowering in the fine sand layer, leading to blockages in the front grouting device. Therefore, in such soil conditions, it is recommended to implement anti-seepage treatment for the pipe joints to effectively prevent frequent blockages. Lastly, the grouting effectiveness was unsatisfactory. The grouting in the sand layer had slow effectiveness and a low success rate. Even by doubling the original grouting parameters, it was challenging to achieve the same results as in soft soil grouting. Therefore, it is suggested to employ multiple rows and grouting operations in

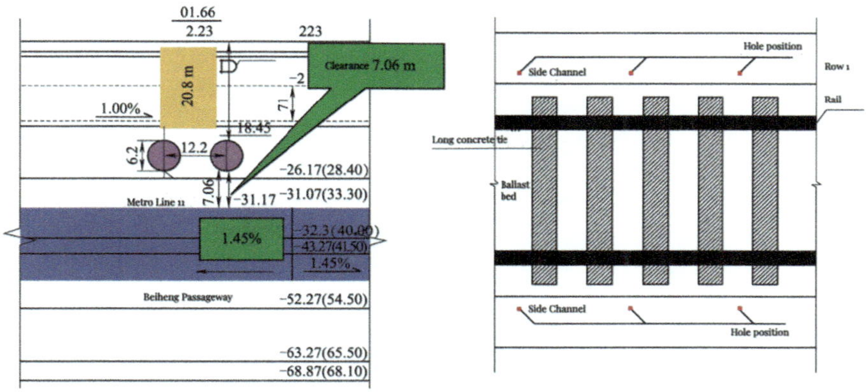

Fig. 8.7 Relative position relationship between Beiheng Passageway and Line 11, and the plan layout of the hole positions (m)

the sand layer grouting, with a single hole grouting volume exceeding 1.5 times the grouting volume used in soft soil grouting.

8.1.3 A Micro-disturbed Grouting of Shanghai Metro Line 11 for a Crossing Area of Beiheng Passageway

8.1.3.1 Background of the Project

The new Beiheng Passageway tunnel crosses the section between "Jiangsu Road Station" and "Longde Road Station" of Line 11. The crossing caused uneven settlement. This is the first time that an extra-large diameter shield tunnel has been constructed beneath an operating metro line, without disrupting the service of Line 11. To ensure the safety of Line 11 during the crossing, emergency measures were taken for controlling potential uneven settlement. Therefore, prior to the crossing construction, pre-drilling was carried out within Line 11 to facilitate potential micro-disturbed grouting. The pre-drilling area covers the up line, ranging from S583 to S622, and the down line, ranging from X568 to X607, totaling 80 rings. Figure 8.7 illustrates the positional relationship between Beiheng Passageway and the Line 11 profile, along with the arrangement of planned boreholes.

8.1.3.2 Construction Process

According to the construction requirements, micro-disturbed grouting was carried out to uplift the tunnel lining with settlement greater than 5 mm after the completion

Fig. 8.8 Photos on the site of the pre-embedded pipes and the second-time grouting boreholes

of the crossing. The grouting area for the up line spanned from S596 to S611, totaling 16 rings, while for the down line, it was from X591 to X601, totaling 11 rings.

The grouting operation to uplift Line 11 began on July 12, 2019, and ended in the early morning of August 3, 2019, with a total of 80 grouting cycles. In the up line, 40 grouting cycles were performed in the S596–S611 ring range (with rings 600, 604, 606, and 608 completing 2 grouting cycles, while the rest completing 1 grouting cycle). In the down line, 40 grouting cycles were performed in the X591–X601 ring range (with rings 592, 596, and 601 completing 2 grouting cycles, and rings 591, 595, and 599 completing 3 grouting cycles).

Figure 8.8 displays the location of the pre-embedded pipes and the second-time grouting boreholes.

8.1.3.3 Construction Effect

Figure 8.9 depicts the cumulative settlement variation curve during the automatic monitoring and grouting stage (from July 12th to August 3rd). Up to August 3rd, the cumulative settlement for the up line was − 1.84 mm, with a cumulative variation of + 7.73 mm. For the down line, the cumulative settlement was − 2.3 mm, with a cumulative variation of + 10.81 mm.

8.1.3.4 Analysis and Conclusion

Based on the observed tunnel settlement evolution, it can be observed that the micro-disturbed grouting technique, through multiple micro-disturbed uplifting operations and continuous settlement treatment of the tunnels, gradually elevate the tunnels

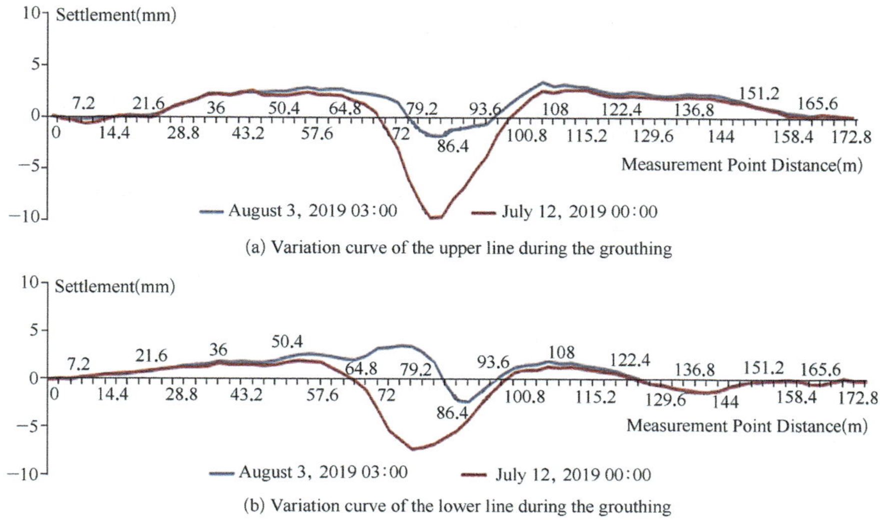

Fig. 8.9 Variation of the tunnel settlement

while minimizing the impact on the metro operation. In this project, with fewer than 2 grouting cycles, the cumulative settlement of the tunnels decreased to less than 3 mm, demonstrating significant effectiveness.

8.2 Engineering Cases of Internal Intension Steel Ring

8.2.1 A Tunnel Structure Repair of Shanghai Metro Line 2 Between "Chuangxin Middle Road Station" and "Huaxia East Road Station"

8.2.1.1 Background of the Project

In the section of Shanghai Metro Line 2 between "Chuangxin Middle Road Station" and "Huaxia East Road Station", the up line faced significant abnormal deformation due to excessive surcharge during metro operations. The details are as follows:

(1) Between the 540th and 620th rings on the up line, there was notable widening of longitudinal seams on the top segments of the tunnel, along with substantial lateral deformation of the tunnel.
(2) Severe water leakage occurred between the 270th and 640th rings on the up line.

(3) Extensive fragmentation of tunnel segments took place between the 400th and 625th rings on the up line.
(4) Bolt and nut detachment was observed on the tunnel segments at the 410th and 581st rings on the up line.

To ensure the safety of the tunnel structure and the normal operation of Metro Line 2, specific technical measures are required for repairing, reinforcing, and strengthening the tunnel lining within this designated range.

According to geological data, the center elevation of the up line tunnel section is approximately − 15.0 to − 16.0m, while the ground surface elevation is about 4m. The thickness of the covering soil is around 16.5 m. The primary soil layers in the tunnel area consist of silt clay layer ④ and gray powdery clay $⑤_1$.

8.2.1.2 Construction Process

Given the evident structural damages and water leakage in the tunnel, and according to repair principles and standards, structural safety repair techniques will be employed for restoration. The structural reinforcement mainly includes: firstly, temporary reinforcement using aramid fabric to prevent further deterioration of the tunnel deformation in the affected area; secondly, internal steel ring support and reinforcement for the problematic segment.

1. The final determined repair scope of the project

   (1) Tunnel segments with significant widening of longitudinal seams and substantial lateral deformation (up and downstream lines, Ring 540 to 620)
   (2) Tunnel segments with severe water leakage (up line, Ring 270 to 640)
   (3) Tunnel segments with substantial fragmentation (up line, Ring 400 to 625)

2. The main contents of reinforcement treatment

   (1) Application of aramid cloth on the top of the segment: for segments with larger longitudinal gap openings, aramid cloth is first applied for reinforcement (Ring 500–600 for the up line, totaling 101 rings). The aramid cloth is divided into three pieces with the handholes as boundaries, and the circumferential arc length is 800 mm. The high temperature resistance index of the aramid cloth is 80 °C. The current situation after applying the aramid cloth in the tunnel is shown in Fig. 8.10.
   (2) Joint grouting for leakage treatment: perform full-range surface ring and longitudinal joint elastic epoxy grouting for Rings 500–600 on the up line. Apply epoxy grout inside the joints.
   (3) Circular steel rings installation: for Ring 500–600 on the up line, a total of 101 rings are reinforced with steel rings, of which 60 rings (Ring 500–559) are semi-ring reinforcement, and 41 rings (Ring 560–600) are full-ring reinforcement. After completing jointing and grouting, install circular steel rings on the inside of the segments as permanent reinforcement for the

Fig. 8.10 Current condition of the tunnel after pasting aramid cloth

tunnel. The steel rings have a thickness of 20 mm (the steel plate corbel thickness is 30 mm) and a width of 850 mm. The average gap between the steel plate and the segment should be controlled within 2 mm. The edge of the steel plate is sealed with elastic epoxy grout, and then rigid epoxy grout (DP40) is injected between the steel rings and the concrete segments. Before laying the steel plate, quick repair mortar cement should be used to seal the four handholes with the lowest horizontal position of the standard block and adjacent block. The current situation after the installation of the steel rings is shown in Fig. 8.11.

(4) Concrete segment crack treatment: generally, leakage is addressed by low-pressure grouting along the joint, using DP40 epoxy resin grout. For leaks with sediment, seal with quick-setting cement, and fill two-component polyurethane grout.

Fig. 8.11 Tunnel reinforcement with steel rings

Fig. 8.12 Appearance of the tunnel after sealing the leaks

(5) Fractured segments treatment: repair severely fractured segments. Use quick repair mortar cement for filling, inject DP40 epoxy resin into the joints through pre-installed grouting pipes to enhance joint bonding, and prevent the filling material detachment.
(6) Back grouting treatment: within the range of Ring 270 on the up and down lines to the adjacent passageway, excluding the 101 rings that were previously reinforced with steel rings, segments with significant leakage are treated with back pressure injection of polyurethane to seal the leaks. The grouting is located at the grouting hole reserved in the standard block (B block), and the grouting pressure is determined based on the actual conditions on-site. Apply 30 kg back grouting per hole, totaling 60 kg per ring for two standard blocks. The appearance of the tunnel after sealing is shown in Fig. 8.12.

8.2.1.3 Construction Effect

(1) The tunnel deformation is under control, and cease to progress, and overall stability has been maintained.
(2) The opening of longitudinal gaps in the tunnel has been well controlled.
(3) Water leakage in the tunnel has greatly improved, with minimal visible seepage or dripping, contributing to noticeable improvement in overall appearance.
(4) Convergence in the tunnel diameter: tunnel diameter convergence has been well controlled through the application of a safe repair technique using internal tension steel rings. (The convergence is within $\pm$ 10 mm, meeting the design requirements, as showed in Fig. 8.13.)

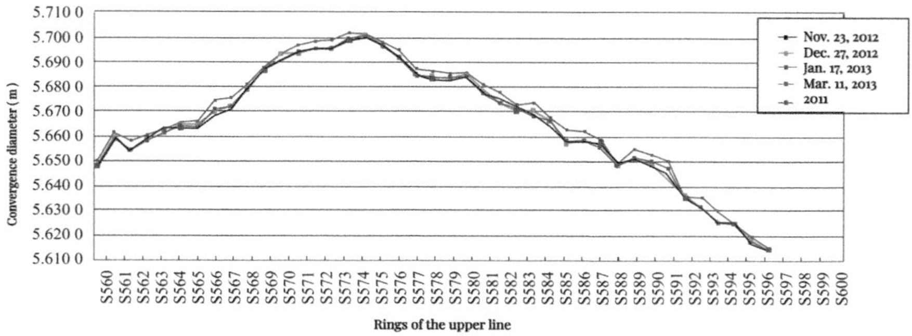

Fig. 8.13 Convergence curve of the tunnel diameter

8.2.2 A Tunnel Repair Between "Dongan Road Station" and "Houtan Station" of Shanghai Metro Line 7

8.2.2.1 Background of the Project

The tunnel section between "Dong'an Road Station" and "Houtan Station" on Shanghai Metro Line 7 is an important Expo route for Shanghai's rail transit. During the operation, the following anomalies were discovered in some tunnel sections:

(1) Severe water leakage in some tunnel sections.
(2) Through further convergence monitoring of the tunnel, significant convergence was found greater than 100 mm.
(3) Some segments suffer from fragmentation.

To ensure the long-term structural safety of these tunnel sections, especially for the normal operation of Shanghai Metro Line 7, specific technical measures need to be taken to reinforce and repair the affected tunnel sections.

8.2.2.2 Construction Process

Based on the analysis and investigation of the structural damage in the tunnel section between "Dong'an Road Station" and "Houtan Station" of Shanghai Metro Line 7, the scope and content of the repair were determined, as shown in Table 8.2.

8.2.2.3 Construction Effect

By reinforcing the tunnel with internal intension steel rings, the convergence of the damaged tunnel has been effectively controlled, contributing to a stable tunnel structure without any water leakage. Analysis of monitoring data after repair for

Table 8.2 Repair contents of the structural damage in the tunnel section between "Dong'an Road Station" and "Houtan Station"

Section	The number of the steel ring (ring)			The number of aramid fiber reinforced polymers (ring) (convergence exceeding 100 mm)
	Non-overlapping joint ring (convergence exceeding 120 mm)	Overlapping joint ring (circumferential misalignment of tunnel lining greater than 20 mm)	Relocation of supporting pipelines, roadbed cutting, and removal of rings	
Down line from Dong'an Road Station to Houtan Station	36 (whole ring) + 74 (semi-ring)	0	110	110

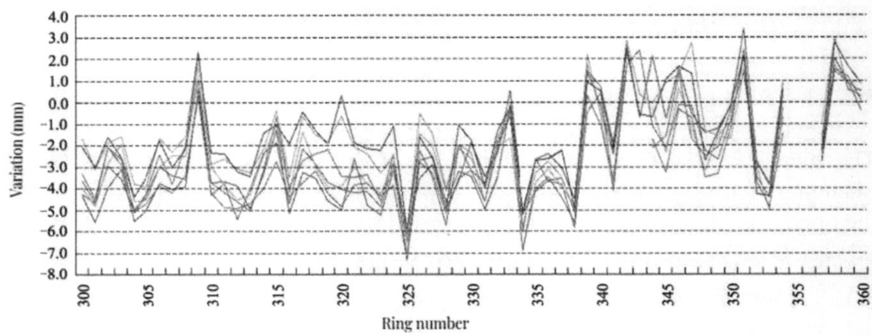

Fig. 8.14 Convergence curve of the tunnel diameter

several months shows that the convergence of the tunnel is controlled within − 8 to 4 mm, meeting the design requirements, as shown in Fig. 8.14.

8.2.3 A Tunnel Repair of Nanjing Metro Line 2 Between "Jinma Road Station" and "Maqun Road Station"

8.2.3.1 Background of the Project

In 2012, the operational tunnel section between "Jinma Road Station" and "Maqun Station" on Nanjing Metro Line 2 encountered problems due to soil settlement on the surface. The following issues were identified:

(1) Significant tunnel convergence was observed through in-situ monitoring, with a maximum of 118.8 mm.

(2) Penetrating cracks appeared in segments.
(3) Noticeable water leakage was present on both sides of the tunnel.

To ensure the long-term structural safety of this tunnel section, particularly for the normal operation of the metro, structural reinforcement of the segments is currently underway.

8.2.3.2 Construction Process

According to the data provided by the owner and the design unit, aramid cloth was applied to a total of 66 rings, circumferential and longitudinal joints sealing was performed on 71 rings, and steel ring reinforcement was carried out on 30 rings.

8.2.3.3 Construction Effect

The convergence of the tunnel has been controlled within − 0.44 to + 0.20 mm. The convergence has been effectively managed, and therefore the tunnel has reached a stable state.

8.2.4 A Tunnel Repair of Wuhan Metro Line 2 Phase One Between "Changgang Road Station" and "Hankou Railway Station"

8.2.4.1 Background of the Project

The section between "Changgang Road Station" and "Hankou Railway Station" of Wuhan Metro Line 2 is a single-line circular tunnel with a segment outer diameter of 6.0 m, inner diameter of 5.4 m, segment thickness of 300 mm, and a ring width of 1.5 m.

In March, 2016, there was a large-scale soil deposition on the ground in the DK3 + 750 ~ DK3 + 945 section of Changhan Tunnel. The deposited soil was 6 m higher than the originally designed cover depth of 21.8 m. This led to significant convergence of certain tunnel segments, with minor cracks appearing at the top segments. Some segments experienced penetrating cracks with a width exceeding 0.2 mm. Additionally, extensive water leakage occurred, and there was some extent of void formation in the roadbed, as shown in Fig. 8.15.

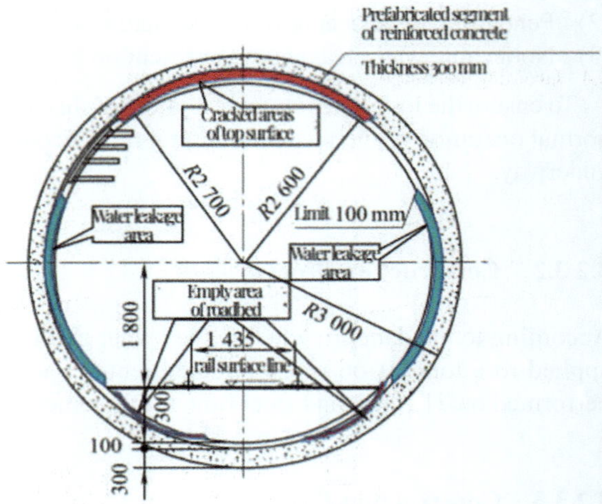

Fig. 8.15 Schematic diagram of the tunnel defects in the section between Changgang Road and Hankou Railway Station (unit: mm)

8.2.4.2 Construction Process

To ensure the safety of metro operation, after thorough discussions with the owners, designers, and relevant experts, it has been decided to prioritize addressing water leakage in the tunnel section. Once the tunnel convergence and settlement stabilize, cracks and segment reinforcements will be carried out. For segments with convergences exceeding 70 mm in absolute value and penetrating cracks wider than 0.2 mm, steel ring lining reinforcement will be implemented. The installation of steel ring lining will be conducted after the tunnel deformation stabilizes and the surrounding environment is under control. Cracks wider than 0.2 mm will be treated with epoxy filling and external aramid cloth application. For extensive leakage, grouting with oil-soluble polyurethane will be employed on the segment walls, while water-soluble polyurethane grouting will be used for drip leakage along the ring and longitudinal joints. As for voids in the roadbed, epoxy filling reinforcement will be performed, as shown in Fig. 8.16.

8.2.4.3 Construction Effect

By implementing internal intension steel rings for reinforcement, the convergence of the damaged tunnel has been effectively controlled, resulting in a stable tunnel structure without any water leakage. During the final stage of the repair treatment in the Changhan section, a re-measurement of the convergence was conducted. The sectional convergence monitoring data indicated that both the left and right lines are within the normal range. The convergence in the left line and right line ranged from − 3.5 to + 5.3 mm, from − 3.8 to + 5.6 mm, respectively. The tunnel convergence has reached a stable state, as shown in Fig. 8.17.

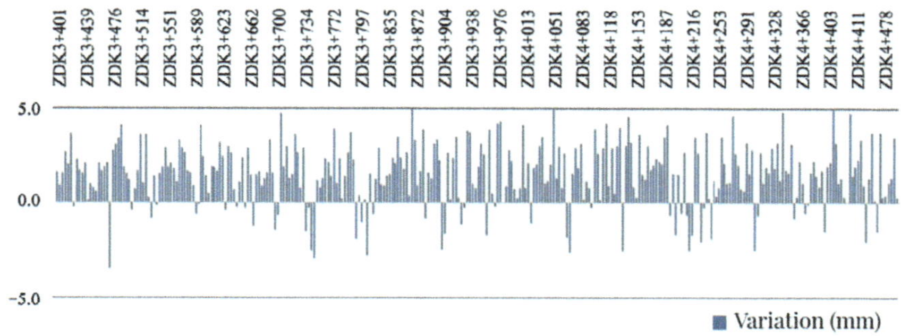

Fig. 8.16 Summary of convergence of monitoring section in the left line of "Changgang Road Station" to "Hankou Railway Station"

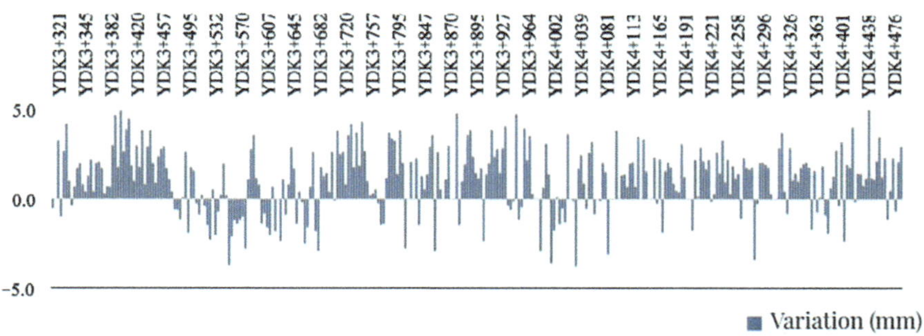

Fig. 8.17 Tunnel convergence in the right line of "Changgang Road Station" to "Hankou Railway Station"

8.3 Engineering Cases of Three-Dimensional Laser Scanning Detection

8.3.1 An Annual Convergence Detection of Mobile Laser Scans of Shanghai Metro Line 2, Line 10, and Line 12

8.3.1.1 Background of the Project

According to the overall plan for the operation and maintenance management of Shanghai metro network, shield tunnel sections require an annual convergence monitoring at least once a year. Based on the monitoring results, it is necessary to promptly detect sections with large convergence and take measures to address the corresponding problems, so as to ensure the safety of metro structure and train operation. As part of the convergence measurement, the main workloads of metro Line 2, 10 and 12 are shown in Table 8.3.

Table 8.3 List of long-term convergence measurement workloads

Number	Metro line	The workloads of the line (ring)
1	Main line of Line 2	4966
2	Eastern extension of Line 2	6283
3	Western extension of Line 2	2064
4	Western-western extension of Line 4	2404
5	Metro Line 10	9880
6	Metro Line 12	5973
7	Western line of Line 12	6546
Total		38,116

Shanghai's *Specifications for Monitoring and Measurement of Urban Rail Transit Structure* stipulates that the accuracy of long-term convergence should not be less than $\pm$ 3.0mm. This project adopted mobile laser scanning detection with Faro330 laser scanner and two Z+ F6012 scanners.

8.3.1.2 Monitoring Process

From 2016 to 2019, three annual laser scans of the whole line were completed, forming its convergence measurement results. The previous measurement processes included instrument calibration, field collection, data processing, quality inspection, storage management of results, etc.

8.3.1.3 Monitoring Effect

The project utilized mobile laser scanning for convergence detection, generating comprehensive convergence and internal structure wall images. This provided essential data for security of approximately one-third of Shanghai's entire rail network. The results were managed using a unified data management platform for monitoring cloud-based structures, facilitating data analysis and retrieval, as shown in Fig. 8.18.

8.3.1.4 Analysis and Conclusion

Mobile laser scanning for tunnels is an innovative testing technique that acquires the geometric coordinates and laser reflectance of the tunnel structure with high density. A single scanning operation can provide information on structural deformation and measurable high-resolution laser images. Compared to traditional testing techniques, it offers comprehensive advantages such as full coverage, high precision, and high efficiency.

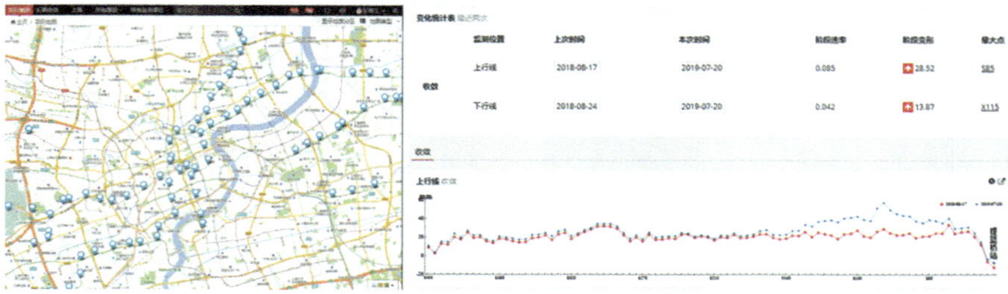

Fig. 8.18 Management of tunnel convergence measurements results and access to the data of a certain section

8.3.2 A Shield Tunnel Convergence Detection of Mobile Laser Scans of All Nanjing Metro Lines

8.3.2.1 Background of the Project

Up to November, 2016, Nanjing Metro operates 6 lines with 121 stations, spanning a total length of 224.4 km. Nanjing's geological conditions are complex, with its metro tunnels crossing multiple geological units and encountering significant variations in soil conditions along the route. The construction experiences of many cities suggest that metro tunnels are prone to deformation during operation due to various factors, including ground and surrounding building loads, nearby construction activities, and train-induced vibrations.

To fully examine the convergence of existing shield tunnels and to provide basic data for subsequent metro maintenance, Nanjing Metro implemented a project for measuring the ring diameter convergence for the entire shield tunnel network. The project covered all shield tunnels of the six lines: Line 1, Line 2, Line 3, Line 10, Line S1, and Line S8. It included 11 single-ring sections for Line 1, 13 for Line 2, 24 single-ring sections plus 1 large-ring section for Line 3, 8 single-ring sections plus 1 large-ring section for Line 10, 5 single-ring sections for Line S1, and 6 single-ring sections for Line S8.

In total, the project measured 144, 425 convergence cross-sections, totaling 178.8732 km for the entire network. Details regarding measured routes, lengths, section counts, total convergence sections, and fieldwork are provided in Table 8.4.

8.3.2.2 Monitoring Process

The field survey of the project started on August 17 and ended on October 25, 2016, and the calculation was completed in December 2016 with the submission of the survey report. The project involved processes such as scheme review, experimental comparison between laser scanning and traditional methods, scanning operations, reliability checks on the results, and data storage.

Table 8.4 Workload of convergence monitoring of all Nanjing metro lines' shield tunnels

Metro line	Completed mileage (km)	The number of passages		The number of sections		The field workload/shift group day
		Single-circle	Large-circle	Single-circle	Large-circle	
Metro Line 1	21.17	11	0	17,645	0	9
Metro Line 2	27.26	13	0	22,718	0	11
Metro Line 3	64.13	24	1	47,851	3354	26
Metro Line 10	22.30	8	1	12,581	3600	12
Airport Line S1	28.27	5	0	23,556	0	11
Ningtian Line S8	15.74	6	0	13,120	0	6
Total	178.8	67	2	137,471	6954	75

The project results were managed by an information system, with each section treated as an individual project. Data collectors uploaded the convergence diameter results and tunnel image data. Besides, the system, equipped with data analysis and reporting tools, facilitated the implementation and management of the project.

The convergence monitoring results for the entire shield tunnel network of Nanjing Metro were divided into 69 sections, as shown in Fig. 8.19.

The convergence diameters and images management and publication of an individual project are illustrated in Fig. 8.20.

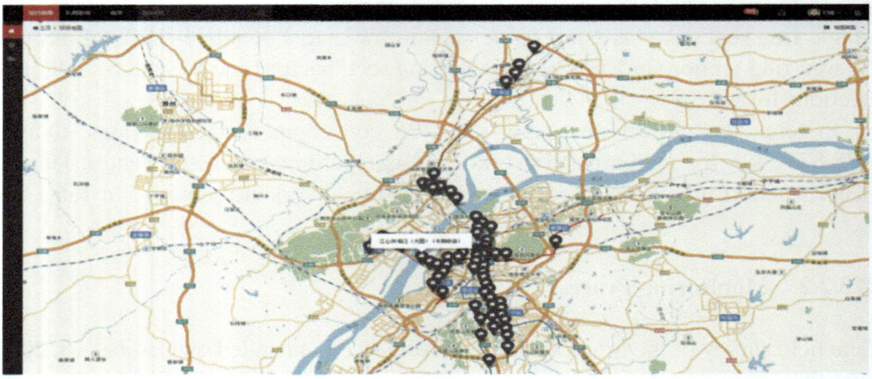

Fig. 8.19 Informatization platform of convergence monitoring of all Nanjing metro lines' shield tunnel

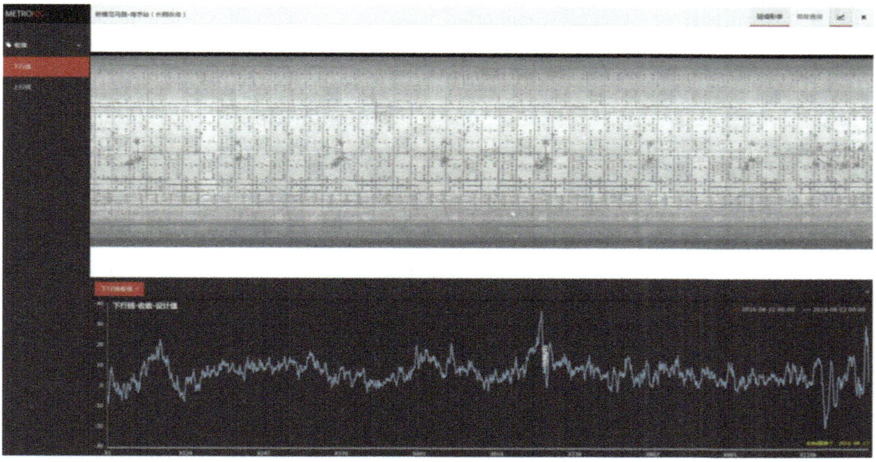

Fig. 8.20 Display of convergence monitoring project management from Nanjing Station to Xinmofan Road Station on Nanjing Metro Line 1

The cloud mapping system has built-in data analysis functions that allow analysis and querying of all convergence diameter results by lines, sections, and ranges of variation.

As demonstrated in Fig. 8.21, the system can also provide information on results data by lines based on the range of variations, and offer related details, such as maximum (minimum) convergence values and workload statistics.

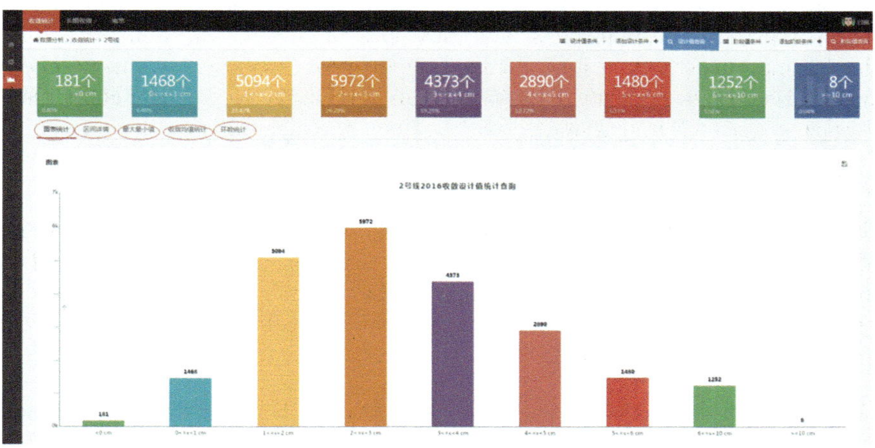

Fig. 8.21 Statistical analysis chart of convergence monitoring results

8.3.2.3 Monitoring Effect

The convergence measurement project for the entire Nanjing Metro network was completed within the designated time using mobile laser scanning technology. The measurement results were more accurate, better than the expected requirement of ± 3 mm, indicating the reliability of the measurement results.

Through the statistical analysis of the measurement results, a total of 34 intervals (single-line) were detected with convergence exceeding 60 mm in the alarm zone, accounting for 25% of the total monitored intervals and 1.1% of the total monitored cross-sections (1624 in total). There were 14 intervals (single-line) with convergence exceeding 80 mm, totaling 209 cross-sections (excluding the 12 rings of the large-circle shield tunnel). These accounted for 10.3% of the total monitored intervals and 0.1% of the total monitored cross-sections.

8.3.2.4 Analysis and Conclusion

The three-dimension laser scanning allows rapid detection of tunnel structural diseases, characterized by high operational efficiency, reliable results, and rich product offerings. It can be considered a leading technology of convergence measurement. The informatization management system of measuring results can be used for analyzing convergence data management and scanning image results, serving as an important decision support system for future metro maintenance.

With the increasing construction activities near metro lines, many more complex factors will cause tunnel deformation. It is necessary to regularly conduct a comprehensive survey of tunnel convergence using convergence monitoring of the mobile laser scanning for the entire network near metro lines, ensuring the latest health information of metro tunnels.

8.3.3 A Tunnel Structure Inspection Using Laser Scanning

Take the example of metro tunnel structure inspection during the construction period of "China-US Trust" site.

8.3.3.1 Background of the Project

The site for this project is located in Hongkou District, bounded by Wusong Road, Suzhou Road, Zhapu Road, and Tiantong Road. The project is called "Sino-US Trust Financial Building Project" and is primarily intended for financial and office use. The project consists of two tower buildings and a commercial podium. The common area of the foundation pit has four basement levels, while the adjacent metro area has two basement levels. The construction of the project started from

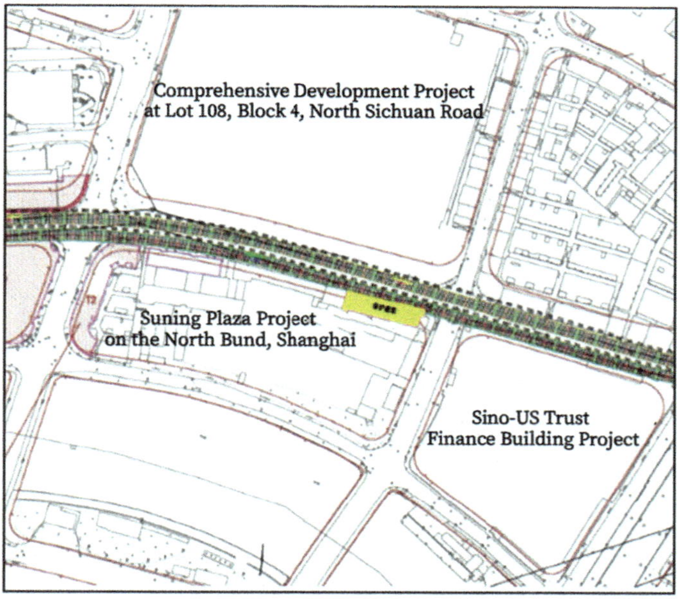

Fig. 8.22 Relative relationship between the foundation pit and metro Line 12

July 2013 to August 2018. The project would influence the tunnel section between "Guokezhongxin Station" and "Tiantong Road Station" on Metro Line 12. The length of the parallel foundation pit and metro tunnel is approximately 110 m, corresponding to Ring 720–810. The closest distance between the basement and the metro tunnel is approximately 10.3 m. The metro up line is situated near the side of the foundation pit. The relative relationship between the foundation pit and the tunnel of Metro Line 12 is illustrated in Fig. 8.22.

8.3.3.2 Inspection Process

Before the construction, at various construction milestones (preliminary/semi-annual, excavation period/quarterly), and after the construction, multiple structural laser scanning inspections were conducted for this project. This process generated a comprehensive and distributed record of the evolution of tunnel convergence, rotation angle between adjacent segments, distribution of defects, and other relevant information throughout the whole process.

8.3.3.3 Inspection Effect

1. Variation of tunnel horizontal diameter

To fully analyze influences of the excavation of the foundation pit on metro tunnels, several scans will be performed before and after the excavation. The monitored horizontal diameter of each scan were shown in Fig. 8.23, depicting six statistical results of horizontal diameters from December 30, 2015 to October 14, 2017. The variation of horizontal diameter can be referred as the tunnel convergence.

"Sino-US Trust Finance Building" includes four excavation zones labeled as A, B, C, and D. Zones B and D directly face the tunnel of Ring 725 to Ring 760, while Zones A and C face tunnel sections of Ring 780 to Ring 810. Excavation dates for each zone are as follows: Zone A from March 21, 2016, to June 4, 2016; Zone C from November 28, 2016, to December 29, 2016; Zone B from February 18, 2017, to May 17, 2017; and Zone D from September 28, 2017, to November 12, 2017.

The picture reveals that large deformation occurred around the horizontal diameter of the segments between Ring 780 and Ring 810 before and after the excavation of the foundation pit in Zones A and C. Similarly, in Zones B and D, significant deformations were observed in the horizontal diameter of the tunnel segments between Ring 725 and Ring 760 before and after excavation.

2. Segment rotation

The high rigidity of the assembling tunnel segments resulted in deformation usually occurred around tunnel joints, namely, the joint dislocation and the rotation between segments. The rotation angle between segments is illustrated in Fig. 8.24 Point J is the connecting point between assembly block l and connecting block l, point C_{pl} is the center of assembly block 1, point C_{11} is the center of connecting block l, and the angle θ between line JC_{p1} and JC_{11} is the rotation angle between assembly block l and connecting block 1.

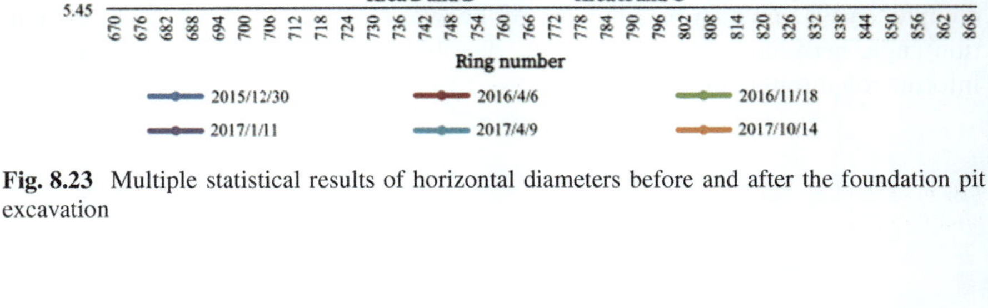

Fig. 8.23 Multiple statistical results of horizontal diameters before and after the foundation pit excavation

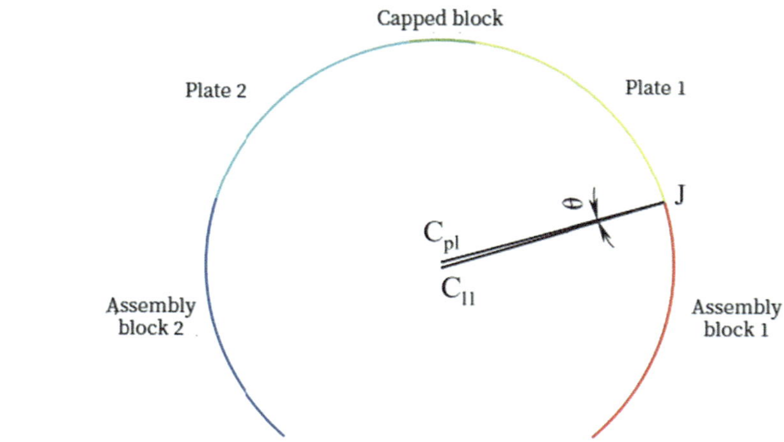

Fig. 8.24 Rotation angle between segments

The project respectively calculated changes of rotation angles between assembly blocks and connecting blocks of the tunnel before and after the excavation of the foundation pit in Zone A. The results are shown in Figs. 8.25 and 8.26.

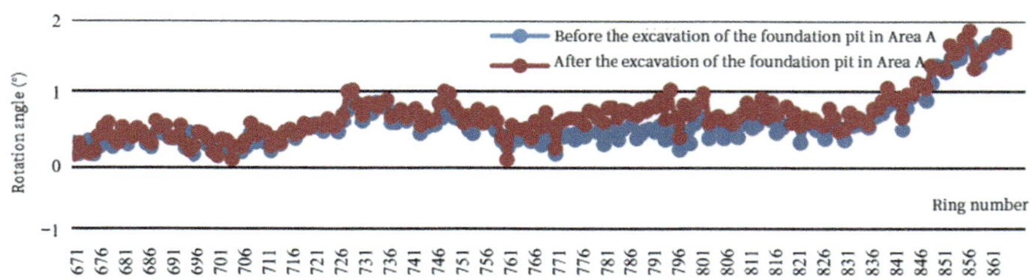

Fig. 8.25 Statistics of rotation angle between assembly block 1 and connecting block 1 before and after foundation pit excavation

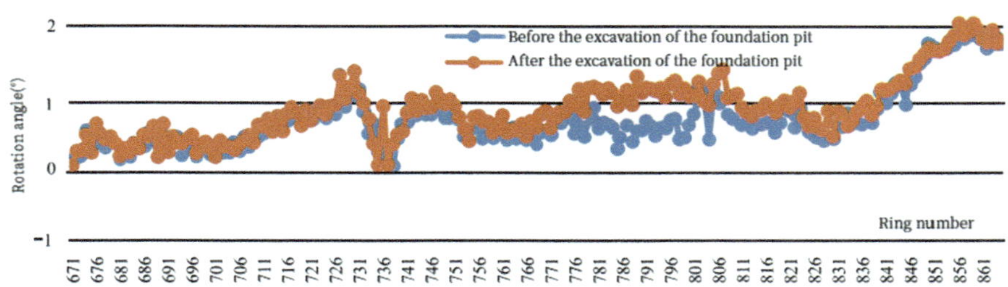

Fig. 8.26 Statistics of rotation angle between assembly block 2 and connecting block 2 before and after foundation pit excavation

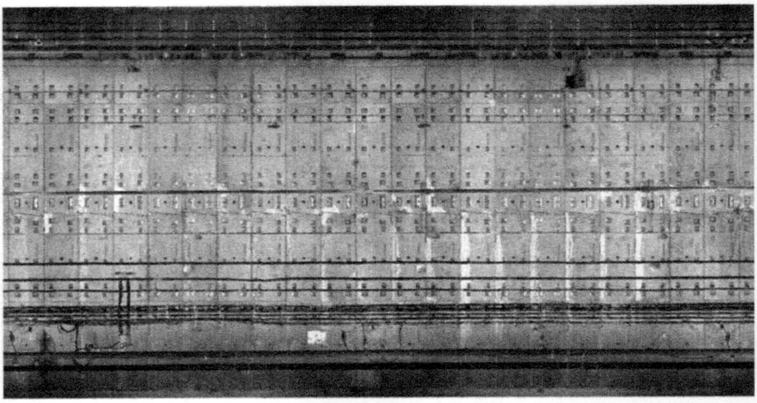

Fig. 8.27 Laser scanning image of Ring 725 to Ring 745 (Oct. 30, 2015)

According to the horizontal diameter statistics of Figs. 8.25 and 8.26, it can be seen that the larger the tunnel deformation, the larger the rotation angle between segments. Due to excavation, the segments between Ring 747 and Ring 817 were severely deformed, accompanied by considerable changes in the rotation angles between the segments.

3. Disease inspection

The reflectivity of the tunnel segments can generate a gray image of tunnel's inner surface. Figures 8.27 and 8.28 show laser scanning images of the tunnel sections obtained by this project on October 30, 2015 and October 26, 2017. There were several water leakage in the tunnel after the excavation of the foundation pit by comparison. Figure 8.29 provides an enlarged view of water leakage of Ring 726 before and after foundation pit excavation. It is evident from the figure that numerous water leakage appeared at the segment joints after the excavation of the foundation pit.

8.3.3.4 Analysis and Conclusion

Three-dimensional laser scanning can be applied for the rapid detection of structural diseases in tunnels, including tunnel convergence, full sections and high-resolution images, inspection of diseases, etc. Featuring high efficiency, low cost, full coverage, and rich information, it can effectively record the whole process of structural deformation and diseases in an all-round way.

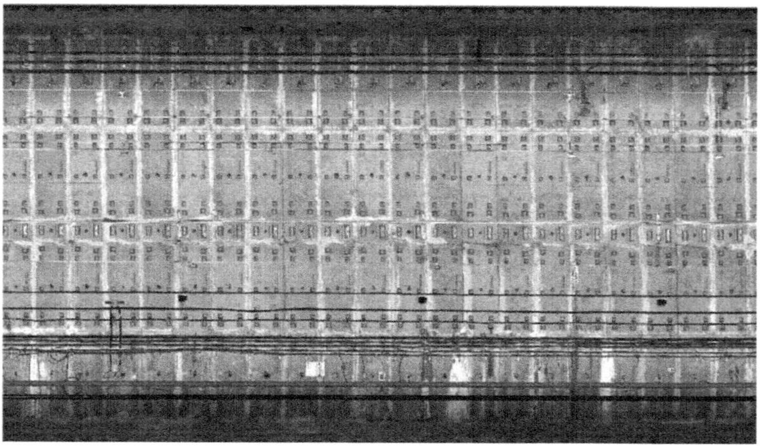

Fig. 8.28 Laser scanning image of Ring 725 to Ring 745 (Oct. 14, 2017)

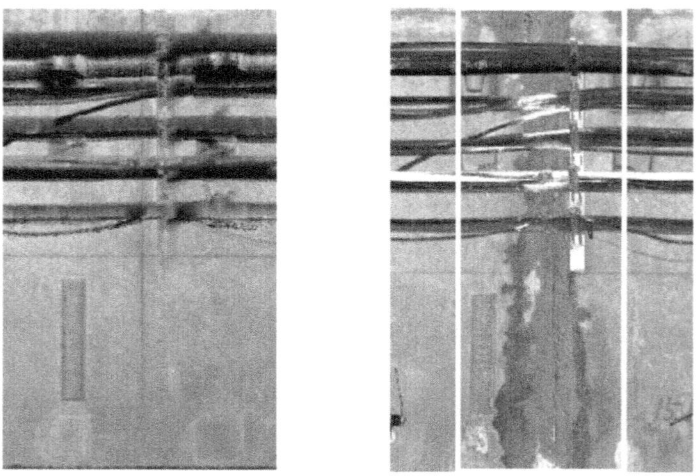

(a) Tunnel image of Ring 726 on Oct. 26, 2015 (b) Tunnel image of Ring 726 on Oct. 14, 2017

Fig. 8.29 Comparison of water leakage in Ring 726 before and after deep excavation

8.4 Engineering Cases of System Platform

8.4.1 Application of the Data Management Platform of Structural Monitoring for Shanghai Metro

8.4.1.1 Background of the Project

Since the opening of Shanghai Metro Line 1 in 1995, 17 metro lines have been put into operation, with a total length of 705 km (including maglev lines). There are 415 operational stations, and the scale of the rail network ranks among the largest in the world. Structural monitoring is a crucial component of metro security. The structural monitoring data include measurement data (such as manual precision measurement data, automatic sensor monitoring data, 3D laser scanning calculation data), image data (3D laser scanning image, UAV aerial image data), other data (working conditions and photos, point layout diagrams), etc.

Data Management Platform of Structural Monitoring for Shanghai Metro was jointly developed by Shanghai Shentong Metro Co., Ltd. and SGIDI Engineering Consulting (GROUP) Co., Ltd. It has been put into service for many years and has achieved good effects.

8.4.1.2 Process of the Project

According to monitoring focus, the metro monitoring projects in the platform are categorized into monitoring and measurement projects, long-term settlement measurement projects, long-term convergence measurement projects and key sections measurement projects.

1. Monitoring and measurement

The monitoring and measuring data include settlement data, convergence data, tilt data, planar displacement data, automatic rangefinder data, and automated static leveling data. The measurement items and measuring point locations are configured based on different levels of projects and actual requirements. Currently, the platform has a total of 287 monitoring and measurement projects, including 158 ongoing projects and 129 completed projects. The quantity and distribution of the projects are shown in Figs. 8.30 and 8.31.

2. Long-term settlement measurement

Long-term settlement measurement involves assessing the settlement of the entire line twice a year, as illustrated in Fig. 8.32. The measurement data primarily consist of the elevation of the measuring points collected by the level. Currently, the platform has long-term settlement data for 181 sections, mainly covering Lines 2, 5, 10, 12, 13, and 16, as shown in Fig. 8.33. Because of different construction timelines for

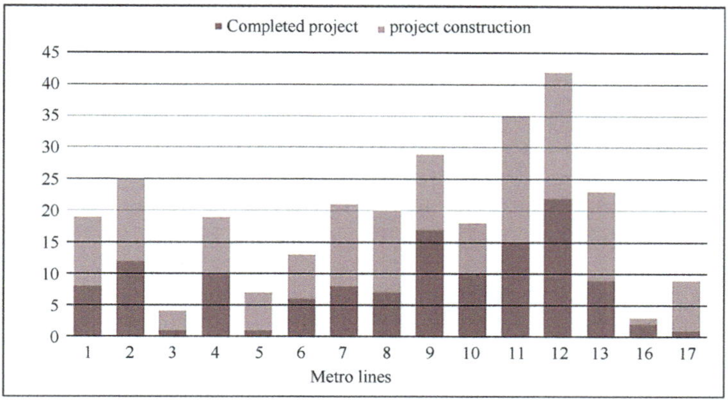

Fig. 8.30 Bar graph of the number of monitoring and measurement projects in Shanghai

Fig. 8.31 Distribution map of monitoring and measurement projects being built in Shanghai

each line, historical data storage ranges from approximately 2007 to 2017, with some sections having historical data traceable back to 1999 (part of Line 2).

The platform can visualize results into increasement of settlement, cumulative settlement, settlement development with time and variation of differential settlement.

3. Long-term convergence measurement

Long-term convergence measurement involves assessing the internal diameter of the entire line once a year. The measurement data include the horizontal diameter of the tunnel collected by the total station and by the calculation of 3D laser scanning point clouds, as well as 3D laser scanning image data and UAV aerial images, as shown in Figs. 8.34, 8.35 and 8.36.

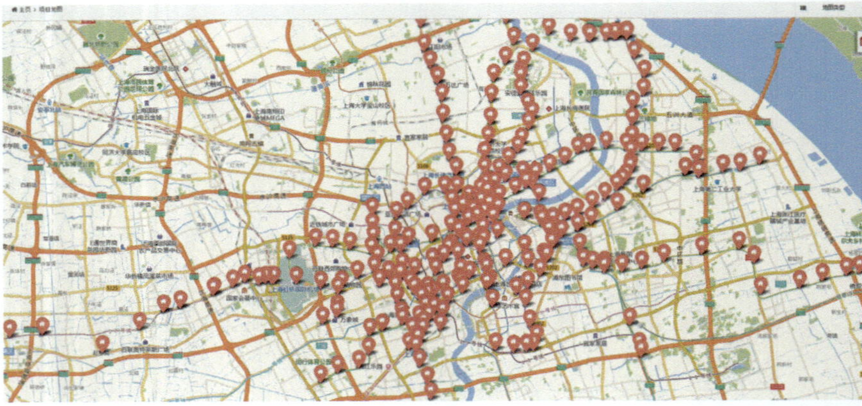

Fig. 8.32 Distribution map of long-term settlement measurement in Shanghai

Fig. 8.33 Long-term settlement of part of Shanghai Metro Line

At present, the platform has 282 sections for long-term convergence, which are mainly operational metro shield tunnel sections. Due to variations in construction timelines for each line, the stored historical data span roughly from 2014 to 2017, with some sections' historical data traceable back to 2011. The platform can visualize results into increasement of convergence, diameter deviation from the design value, cumulative convergence, evolution of tunnel convergence with time. Additionally, the platform can release 3D laser scanning images and UAV aerial images stored in the platform at different periods.

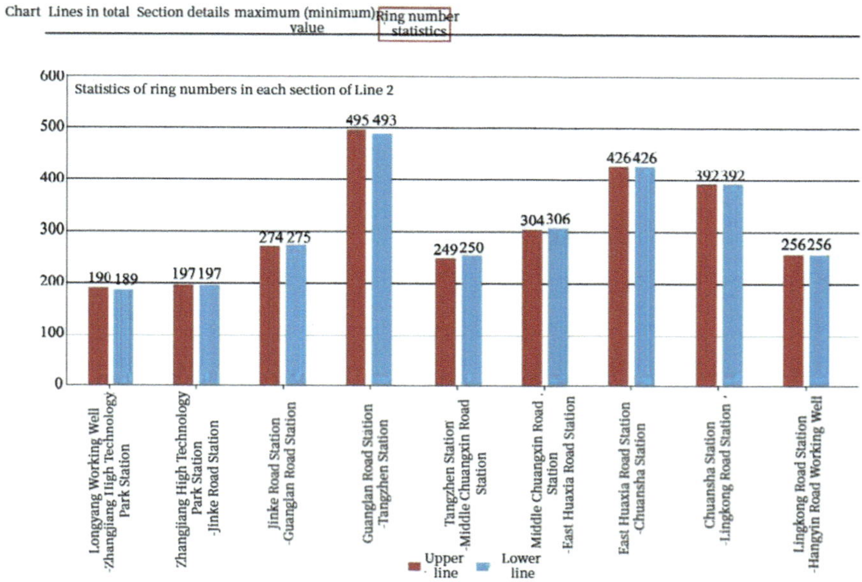

Fig. 8.34 Long-term convergence of part of Shanghai Metro Line

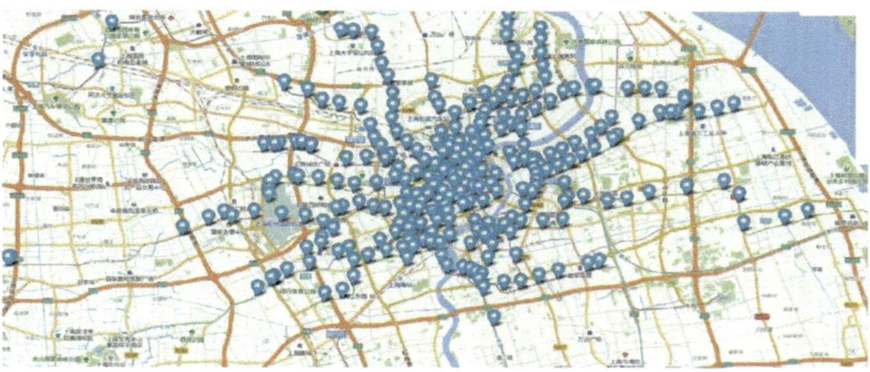

Fig. 8.35 Distribution map of long-term convergence measurement

4. Key sections measurement

Key section measurement refers to regular measurements in the critical areas of the metro that require close attention. The measurement data mainly includes the elevation of survey points collected by the level instrument and the tunnel's horizontal diameter data collected by the total station. Different measurement items can be selected according to the actual measurement needs, as shown in Figs. 8.37 and 8.38.

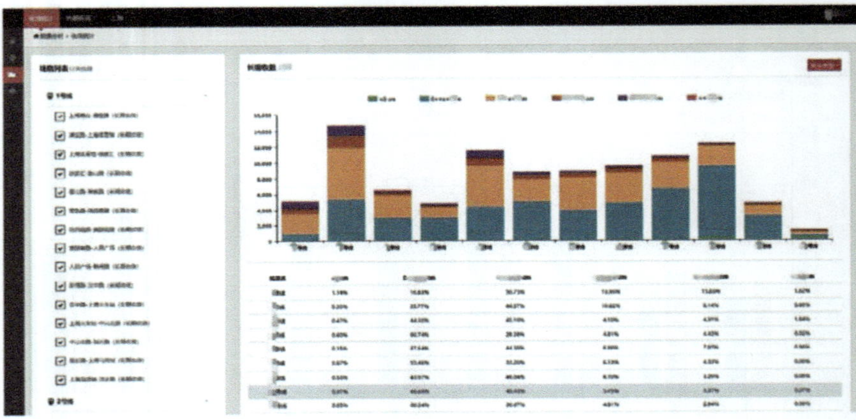

Fig. 8.36 Bar graph of long-term convergence in Shanghai

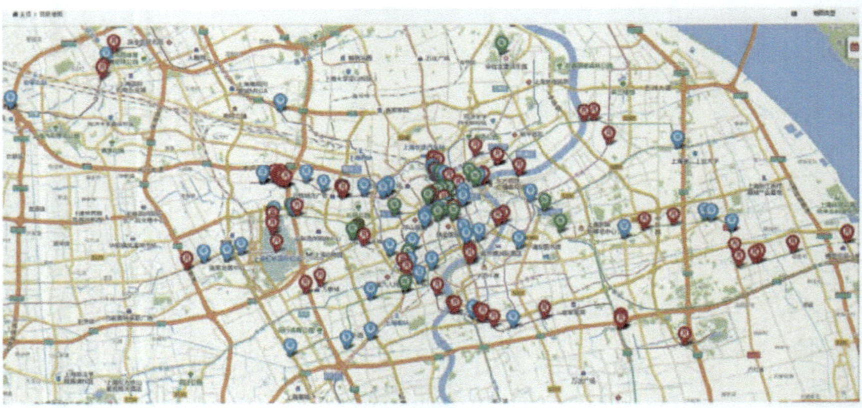

Fig. 8.37 Distribution map of key section measurement in Shanghai

Currently, the platform comprises 111 key sections, mainly including Line 1, Line 10, Line 12, and Line 16. Each key section has a different initial monitoring time, monitoring frequency, and stored data volume. For example, the work shaft from Lingkong Road to Yuandong Avenue, an extension of Line 2, started monthly monitoring in 2013 and changed to a weekly monitoring schedule in January 2017. Up to November 2017, the platform has documented a total of 76 sets of measurement data. The main visualized results on the platform include increasement of settlement, cumulative settlement, development of tunnel settlement with time, and differential settlement evolution. Increasement of tunnel convergence, diameter deviation from design values, cumulative convergence, development of tunnel convergence with time are also provided.

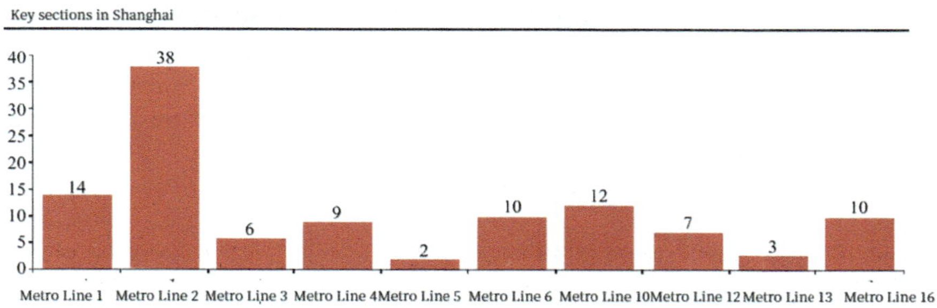

Fig. 8.38 Bar graph of key sections measurement in Shanghai

5. Safety evaluation of tunnel structures

Based on the integration of multi-source data on the platform, attribute recognition method and Hidden Markov Model (HMM) method are used to assess the current safety state of tunnel structures. The first method generates data forms for assessment derived from the multi-source data and performs calculations through algorithm programs. The Hidden Markov Evaluation method is directly embedded in the risk assessment module of the system platform, enabling real-time assessment of monitored projects.

A deep excavation project in Xujiahui, Shanghai is adjacent to Metro Lines 9 and 11. The excavated area of the project exceeds 40,000 m^2 and employs excavation construction in different pits. The average excavation depth of the major pit is around 31.5 m, with the maximum depth reaching approximately 34.5 m (Fig. 8.39). The site is situated on the geological stratum of ancient river channels in Shanghai, with the first and second confined aquifers connected below. The construction of this project poses significant risks.

In Zone 4-3, the excavation has been completed, and during the excavation of the fifth layer of soil in Zone 4-1, the structural safety of the tunnel within the affected construction zone was evaluated. The evaluation results are shown in Fig. 8.40.

8.4.1.3 Construction Effect

The data management platform has supported the comprehensive management and comparative analysis of multi-source sensing and diverse sensing data. Furthermore, real-time assessment of tunnel structure safety based on structural deformation ensures a more comprehensive safeguard, thereby contributing to the secure operation of the metro.

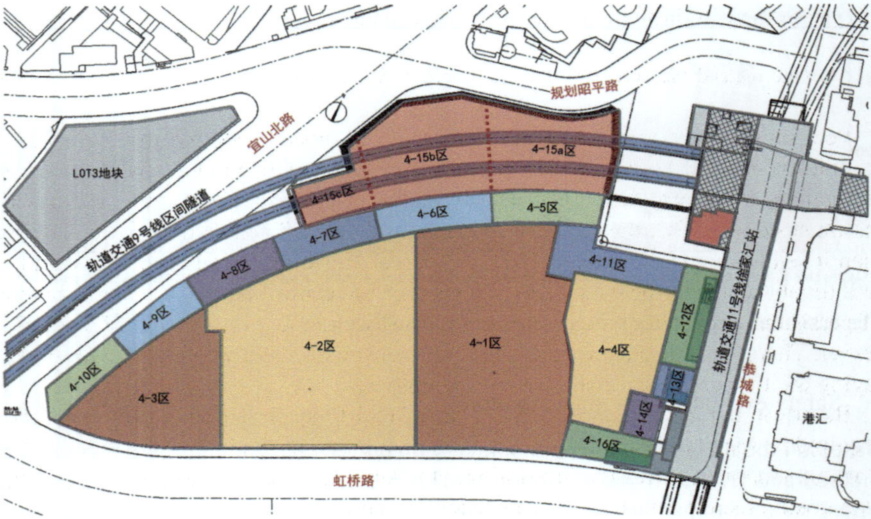

Fig. 8.39 Plan and cross section of the deep excavation project

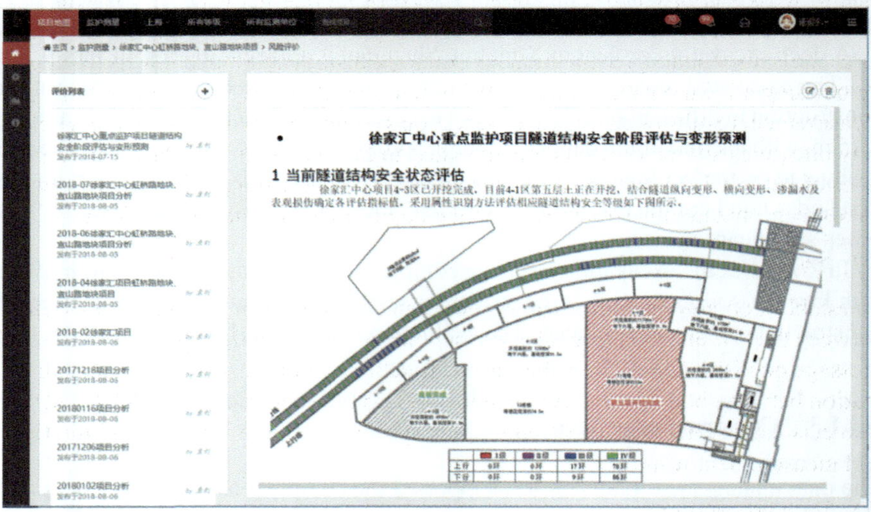

Fig. 8.40 Evaluation results of adjacent tunnel structure during construction

8.4.1.4 Analysis and Conclusion

The application of the monitoring data platform in metro tunnels and other types of projects indicates that the data is transmitted, stored, managed, displayed, and used for decision support through the platform. This ensures efficient and accurate utilization of data, thus improving the operational efficiency of metro tunnel monitoring.

8.4.2 An Application of the Patrol Platform of Dongguan Metro Protection Zone

8.4.2.1 Background of the Project

In the urban rail construction plan of Dongguan, 7 metro lines pass through 32 districts, with a total length of 307.8 km and 84 stations. Among them, Metro Line 2 has been built and put into use, hence this project platform deploys this line. At the same time, Dongguan is situated in the southwest of the Boluo Fault and the Dongguan fault basin on the edge of the Luofu Mountain. The terrain is higher in the southeast and lower in the northwest, with complicated geological conditions. However, current inspection management modes, law enforcement and evidence collection methods, as well as on-site control and management, are evidently insufficient to meet the current and future needs of urban metro protection. How to make full use of advanced informatization technologies to conduct meticulous inspections of the controlled protected areas of the metro network, to effectively supervise and manage the construction, and to ensure the operational safety of the metro tunnels have become important issues to be solved urgently in the inspection and law enforcement of controlled protected areas. Therefore, it is necessary to establish an information-based and intelligent inspection and management platform of protected areas that can technically support scientific monitoring and fine management of Dongguan metro tunnel inspection. The platform includes integrated information management modules, project management modules, inspection management modules, mobile terminals, three-dimensional geological information management modules and system management modules.

The inspection platform for protected areas of Dongguan Metro is mainly divided into two subsystems—the inspection management platform of controlled protected areas (WEB) and the mobile handheld terminal for intelligent inspection (Android). This platform provides a paperless inspection management system for the whole process along the metro protected areas. By embedding intelligent handheld inspection terminals based on GIS maps, GPS positioning modules, and mobile communication technology, it achieves dynamic inspections throughout the entire process of rail transit surface inspections, as shown in Fig. 8.41.

8.4.2.2 Implementation Process

To ensure the universality and availability of platform data, its data processing mainly includes the following components. The data acquisition services at the bottom layer

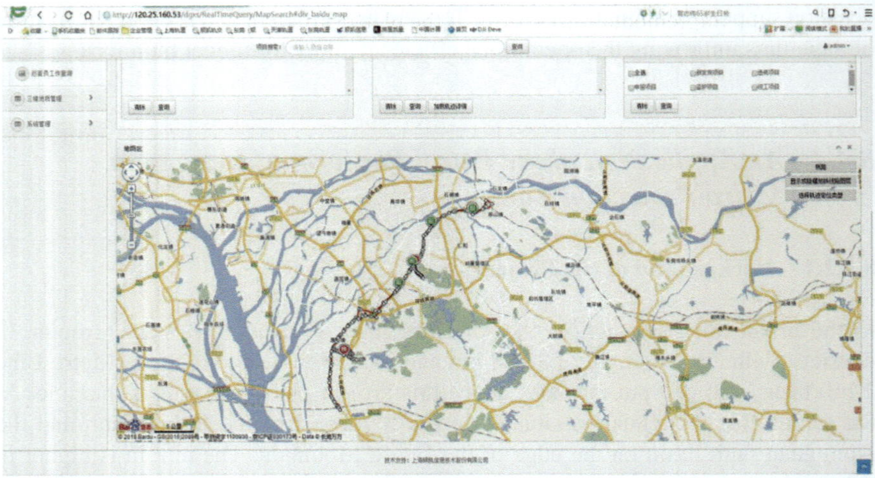

Fig. 8.41 Protected area of Dongguan Metro Line 2

involve activities such as data collection, cleaning, transformation, loading, etc. The data integration services at the middle layer focus on standardizing and organizing the dispersed and various data and records, performing aggregation processing, and providing a shared library of standardized, comprehensive, and dynamic data information (macro or case-level comprehensive business views). The data utilization services at the upper layer mainly include directory access management (information resource positioning, service retrieval, shared database), information utilization management (comprehensive queries, report statistics, online analysis and dynamic alerts, information exploration and auxiliary decision-making analysis), information publishing management, information portals, etc., as shown in Fig. 8.42.

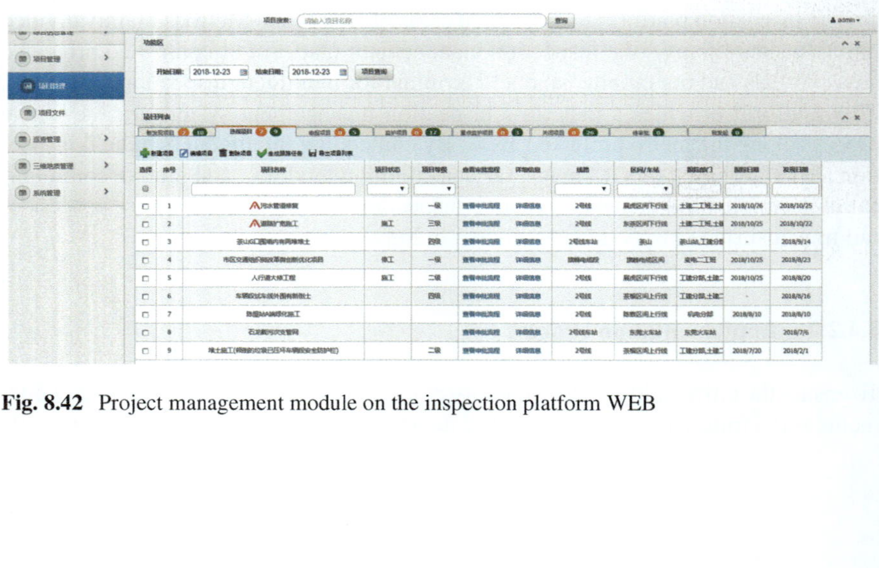

Fig. 8.42 Project management module on the inspection platform WEB

During the implementation process of the platform, the first step is to assign tasks on the WEB end. Once the inspection tasks are assigned, each task order needs to be refined, typically requiring input of the projects and other relevant information within the inspection area. After the task refinement is completed, administrators need to review the refined tasks and provide opinions. When they log in to the handheld terminal which can display all the tasks of this inspection cycle, they can click the date button of a certain day, and view the detailed task categories including routine, tracking and self-built tasks. In the task list, inspectors click the "inspection" button in the upper-right corner of the interface to enter the map interface, where they can see the current task's station, station protected area, section route, and the projects within the section. Click the "Begin to inspect" button at the bottom of the interface to start the job. Subsequently, operations such as "new project", "self-built task", and "report of the main situation" can be completed. At the same time, the movement will be uploaded to the server at a frequency of 20/s, as shown in Fig. 8.43.

Business collaboration services primarily focus on the management of processes, data, and business rules. They also support diverse, flexible, and efficient interactions ensuring comprehensive and effective communication among various business systems and operators. This can accelerate the business process, facilitate business path management, and enhance operational efficiency, service quality, and service levels.

Fig. 8.43 Intelligent inspection mobile handheld terminal

8.4.2.3 Implementation Effect

The inspection platform for protected areas of Dongguan Metro has implemented real-time management processes, sophisticated information retrieval systems, accurate real-time positioning technology and efficient handheld mobile devices. It has generated nearly 200,000 trajectory data points, completed more than 300 tasks, inspected over 1500 sections and 360 shifts, and managed more than 160 projects.

8.4.2.4 Analysis and Conclusion

The inspection platform for protected areas of Dongguan Metro has proved an integrated workflow for urban metro line inspection and project tracking. It establishes a meticulous process from task assignment, refinement, release, and feedback to the query of project distribution. This ensures that each project process is searchable, traceable, and manageable, fully achieving a paperless office environment. With the accurate process recording and collection of the system, users can intuitively retrieve the relevant information from any direction. Real-time tracking of on-site ground inspectors' movements is also available. The system lays a strong basis for users to query relevant information as well as effective supervision mechanism.

8.4.3 An Application of Remote Monitoring for Foundation Pit Metro Structure of "Hengda Center" in Shenzhen

8.4.3.1 Background of the Project

The project site of Shenzhen Hengda Center is located at Baishizhou, Nanshan District, Shenzhen, at the southeast corner of the intersection of Baishi 4th Road and Shenwan 3rd Road, covering a total area of 10,376 m^2. The project plans to construct a super high building (72 floors) with an approximate height of 400 m above ground and a 6-storey basement. The excavation depth of the foundation pit is about 39.05 m and 42.35 m with a rectangular shape. The foundation pit support spans approximately 370 m, covering a total excavated area of about 8633 m^2.

Throughout the construction of the foundation pit, there is enhanced automatic monitoring of adjacent metro structures to assess the deep excavation impact. It also provides timely and accurate warnings for potential hazards or accidents that may threaten the safety of the surrounding environment. This proactive approach enables the prompt implementation of effective measures to eliminate risks and prevent accidents.

Table 8.5 Statistics of the scope of each monitoring project

Monitoring projects	Metro line	Measurement range of monitoring (mileage)
Autonomous measuring robots (vertical displacement of crowns, vertical and horizontal displacement of rail track, and tunnel convergence)	Up and down line of Line 9	DKl + 118 ~ DKl + 398
	Down line of Line 11	DK8 + 580 ~ DK8 + 850
Autonomous measuring robots (vertical displacement of crowns, and tunnel convergence)	Up line of Line 11	DK8 + 580 ~ DK8 + 850
Autonomous measuring robots (vertical and horizontal displacement of rail track)	Western line of Line 11	DK8 + 580 ~ DK8 + 877
Static leveling autonomous measurement of vertical displacement	Western line of Line 11	DK8 + 580 ~ DK8 + 877
	Western line of Line 9	DKl + 118 ~ DKl + 398
Autonomous monitoring of distance measurement equipment (tunnel convergence)	Up line of Line 1	DK8 + 580 ~ DK8 + 850

8.4.3.2 Implementation Process

According to the engineering design drawings, the monitoring targets for this project focus on four tunnels, covering both the up lines and down lines of Metro Line 9 and Line 11. The tunnel monitoring lengths are 120 m right to the deep excavation area, extending 50 m to the west (the up line of Line 11 extends about 77 m, of which the station section is about 27 m and the tunnel section is 50 m), and extending 100 m to the east. The total monitoring range is 270 m (the up line of Line 11 is about 297 m). The mileage range for Line 9 and Line 11 is about DKI + 118 ~ DKI + 398 and DK8 + 580 ~ DK8 + 850 (the range for Line 11 up line is about DK8 + 850 ~ DK8 + 877). Specific monitoring projects are shown in Table 8.5 (Fig. 8.44).

8.4.3.3 Implementation Effect

The project is integrated into a remote monitoring platform, and the statistics of working conditions and deformations are shown in Fig. 8.45. The tunnel deformation can be real-time detected during the deep excavation.

Fig. 8.44 Monitoring points of Line 11 up lines

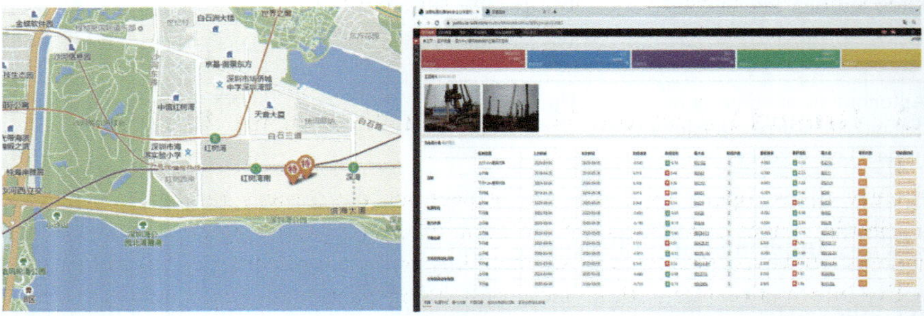

Fig. 8.45 Online monitoring platform for deployment and command delivery

8.4.3.4 Analysis and Conclusion

The implementation of the online remote monitoring platform guarantees that the safety state of Metro Line 9 and Line 11 is controlled in the entire process. Functioning as the "eyes" to monitor metro structure deformations during excavation, the platform promptly issues alerts when monitoring anomalies arise and also accumulate project experiences, serving as a reference for similar projects in the future.

GPSR Compliance

The European Union's (EU) General Product Safety Regulation (GPSR) is a set of rules that requires consumer products to be safe and our obligations to ensure this.

If you have any concerns about our products, you can contact us on ProductSafety@springernature.com

In case Publisher is established outside the EU, the EU authorized representative is:

Springer Nature Customer Service Center GmbH
Europaplatz 3
69115 Heidelberg, Germany

Batch number: 10584688

Printed by Printforce, the Netherlands